PENGUIN BOOKS

AN IMAGINED WORLD

June Goodfield, Senior Research Associate at Rockefeller University, was born in Stratford-on-Avon, England, and received her Ph.D. in the History and Philosophy of Science at the University of Leeds. She has taught and lectured at American universities since 1968. In 1977 she was the Phi Beta Kappa speaker at the annual meeting of the American Association for the Advancement of Science and is currently Adjunct Professor at Cornell University Medical College. She has also written and directed a number of scientific films. Deeply committed to the public understanding of science, June Goodfield has spent her entire career working at the interface between science and the humanities. She is the author of *The Siege of Cancer*, *Playing God: Genetic Engineering and the Manipulation of Life*, *Reflections on Science and the Media*, and the novel *Courier to Peking*.

AN IMAGINED WORLD

A STORY OF SCIENTIFIC DISCOVERY

by June Goodfield

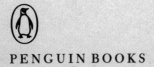

PENGUIN BOOKS

Penguin Books Ltd, Harmondsworth,
Middlesex, England
Penguin Books, 625 Madison Avenue,
New York, New York 10022, U.S.A.
Penguin Books Australia Ltd, Ringwood,
Victoria, Australia
Penguin Books Canada Limited, 2801 John Street,
Markham, Ontario, Canada L3R 1B4
Penguin Books (N.Z.) Ltd, 182–190 Wairau Road,
Auckland 10, New Zealand

First published in the United States of America by
Harper & Row, Publishers, Inc., 1981
Published in Penguin Books by arrangement with
Harper & Row, Publishers, Inc., 1982

LIBRARY OF CONGRESS CATALOGING IN PUBLICATION DATA
Goodfield, G. J. (G. June), 1927–
    An imagined world.
    Reprint. Originally published: New York: Harper &
Row, 1981.
    1. Immunology—Research.   2. Medical research.
I. Title.
QR182.5.G66 1982   616.07'9'072   82-287
ISBN 0 14 00.6204 1   AACR2

Printed in the United States of America by
Offset Paperback Mfrs., Inc., Dallas, Pennsylvania
Set in Caledonia and Melior

For
our three friends
M.E.H. (1923–1978), M.M. (1901–1978), A.Z. (1915–1977)

CONTENTS

PREFACE TO THE
PENGUIN EDITION

It is now over seven years since I first met the scientists whose processes of discovery this book records. The appearance of this paperback edition provides me with a welcome opportunity to answer one question that I've been frequently asked ever since the book first appeared.

Did these scientists march bravely into a cul-de-sac, or was a genuinely new field opened up? It is good to be able to say that a new field does indeed appear to have opened up, and more about this will be found in the Appendix to this edition.

J.G.

New York
1982

PROLOGUE

A letter to the author from the scientist.

July 28, 1975

You want to know what science is? I will tell you. I'll tell you what I did today. After getting in through a corridor of smells, I got myself a little corkboard and the cages of mice. Cages and cages of mice were injected, controls and all; the injected ones with one shot on Sunday. I handled surgical instruments, petri dishes, sterile bottles of medium, sterile bottles for lymph-node cell suspensions. And this is what science is made of.

Oh, it is fine, the great idea, and the best of creative activity, but the day-by-day brickwork is mouse bites which sometimes draw blood; mouse stinks (atoms may not stink, but mice do); the irritating noise of the deep freeze that has gone wrong; sterile bottles, millions of them, washed and autoclaved by the washing-up ladies. Petri dishes, hypodermics, pipettes, packed in some dull chain manipulated by people I do not know; sterile syringes made in some anonymous mass-production system, by people who are not told what the hell I'm going to do with them! They—the technicians, the photographers, and Mrs. Wiggins who washes up our glassware. She looks over sixty and looks as though she has always looked over sixty. She makes up the fantastic infrastructure that we take for granted, that makes it all possible and easy for me.

Without this infrastructure these days there is very little a scientist can do. And if your book manages to tell people just that, to me that is enough. If the man or the woman in the chain who cleans the floor, or who packs the syringe or the needle, hears from your book that by the time it got into the dustbin in this disposable era, their effort wasn't altogether a waste, then your book is worth writing.

Yours,
Anna

The following report on the work of Anna and her colleagues is by way of being a gloss on her letter.

J.G.

The Rockefeller University
New York
1975–1980

PART I

So, if you think that what I say is true,
then by all means agree with me; otherwise you
must use all your resources of logic and argument
to refute me. Make sure I don't deceive you
into sharing my own prejudices and then fly
away, leaving my sting behind like a bee.

PLATO,
The Trial and Execution of Socrates

CHAPTER 1

BETWEEN SOLITUDE
AND SOLIDARITY

Forgive me
If when I want to tell my life,
What I recount is soil.
Such is the Earth.
When it grows in your blood,
You grow.
If it dies in your blood,
You die.

PABLO NERUDA, "Aun" (1969)

To understand the importance of Mrs. Wiggins and appreciate her efforts, it is best to go into the laboratory late at night. The janitors have long since been and gone, erasing the traces of the day's activity by the simple act of sweeping the floor and carrying out the garbage of science: scrawled notes, crunched-up diagrams, plastic coffee cups, wooden spatulas. The contents of the wastepaper basket are just as revealing as Henry Kissinger's ever were.

The floor may have been mopped but the benches are left untouched. The sinks will either be full of dirty glassware—to be dealt with by Mrs. Wiggins in the morning—or empty because the boys have already been in and carried the clutter away to the washer and the autoclave. There is not too much washing up anyway in these disposable days, and thus one traditional ritual of the scientist has vanished from the scene. There was a time, before the advent of technicians en masse, when collecting the dirty glassware and washing it at the end of the day was an ingrained habit of scientists, along with blowing their own glass pipettes and making up the biological solutions. But not now.

But by 11:00 P.M., all the Mrs. Wigginses have passed through and smoothed out the wrinkles of the earlier hours. The rooms are cleaner, the tidiness an orderly chaos, thoroughly familiar to those

who work with it, because they have created both the order and the
chaos and know just where to place their hands so as to find the
stopwatch or the counters—given, of course, that no one else has removed them from their rightful places. The army of anonymous
people who provide the infrastructure for science is thus as silent
and remote as the scientists themselves. But they have left behind
another army: serried rows of beakers, companies of flasks, battalions of tall cylinders, each topped with a silver-papered hat. The
clean glassware in the cupboard and the sterile instruments in the
drawers form the infantry of science, and all it takes to get them under the scientist's orders is an impersonal requisition form, an administrative signature or two, and, of course, the money to pay for it
all. In Illinois or Ohio, the glassware is molded and packed; from
Maine or Minnesota, the sera and the chemical solutions are bottled
and shipped; from California or Washington or London, the check
that makes up the grant that underwrites the scientist who creates
the edifice is signed and sent on its way, by a bureaucrat or a millionaire. Whether microscopes or Magic Markers, centrifuges or
cells, patrons or administrators, the infrastructure of science is a vast
array of people and things—in every respect an industry many orders of magnitude larger than the industry of science itself. The end
result: the indifferent movement that without a moment's thought
casually takes down a beaker from a shelf and marks the start of yet
another experiment.

I first met Anna Brito in February 1975, in the home of a scientist friend. Passing through New York, I had been invited for family
supper to meet two colleagues, and when I went into the sitting
room, I saw two women, the shorter one young, sturdily built,
round-faced, with black curly hair. She looked Mediterranean, and
she spoke with a marked Mediterranean accent. The other was
clearly American. Since these two were scientists and I was writing
about science, we were very soon exchanging politenesses about our
respective jobs. Then we began to dig deeper into ideas, and suddenly, without knowing how we got there, we were in a furious argument about individuality in science. Even more suddenly we
were deep in an argument about scientific recognition. Anna, a
newcomer to America from Portugal via a British university, spoke
of a friend who, she felt, had never received the recognition she de-

served, and what a tragedy that was. With the coldness of a detached observer, I immediately countered that lack of recognition didn't really matter because science is such a communal enterprise that its progress is assured, irrespective of personal recognition. I trotted all this out easily enough, although at heart I believe that individuality *is* of great consequence and significance in science. How it is ultimately expressed is, in fact, a fascinating problem, still virtually unexamined. Anna was quickly on the defensive, although she in turn believes not only in the fact of scientific anonymity and its strength but in its desirability. Thus each of us became devil's advocate, arguing for positions we did not in fact hold.

Nonetheless, within an hour we were, all of us, deep into science. At that time our hostess was interested in the possibility that viruses cause Hodgkin's disease. Anna was puzzling about the same disease for another reason: In patients with Hodgkin's disease, the lymphocytes—the white corpuscles of the blood—gradually disappear from the peripheral blood circulation. Why? Do they vanish completely or are they just hiding somewhere? Soon we were all on the floor surrounded by diagrams and quickly scrawled graphs, looking for a match between the two sets of ideas.

It was not very difficult to find one. Hodgkin's disease shows itself first by fever and all the classic symptoms of a viral infection. It was possible to imagine a situation where, for some reason or other, the lymphocytes were trapped in some organ of the body (Anna thought it might well be the spleen). But wherever they were trapped, the rest of the body was deprived as a consequence of the very cells it needed for self-protection. Since the lymphocytes were not where they ought to be, invading viruses would have the field to themselves, and this would explain not only why the symptoms of a viral infection were so striking but also why the disease was difficult to banish completely. Moreover, if Anna's hypothesis was right, it followed that researchers had been looking to the wrong cause for those deficiencies of immunity that go with the disease, and it also would explain why orthodox treatment might not be totally effective: in order for the body to mount a proper defense against the infection, the lymphocytes would have to be released back into circulation, and in order for *that* to happen, the trap imprisoning them would have to be sprung. Perhaps, she proposed, the root of the whole problem was a genetic defect that showed itself as a change in the makeup of molecules on the lymphocyte's surface. This change could in turn provoke interactions between the lymphocytes and other tissues which could lead to the cell's being held up. In the same way as traffic gets backed up at an intersection because of an accident, so too the lymphocytes might be backed up by cellular circumstances. On the other hand, perhaps interactions between cells themselves were the whole problem, and genetic defects of white blood corpuscles had nothing to do with it. But, Anna insisted, this was all mere fun at this stage—pure, unadulterated speculation. She was on sabbatical leave from a British university to spend six months at an American institute for cancer research, and she hoped to answer one very simple clinical question: In Hodgkin's disease, do the lymphocytes get trapped in the spleen or do they not?

At that time I was wandering around with quivering antennae, wondering whether I would ever meet a scientist whose work I would want to monitor and who would allow me to do so. By the end of those first hours I knew that I wanted to follow Anna Brito's thought and work. I sensed that she was probably at a stage where doors were beginning to open, and the intellectual challenge with its clinical implications intrigued me. Furthermore, she was articulate, amusing, and somewhat different from the majority of scientists I had met before. We agreed to meet again, and one day I asked if I might bring a tape recorder along.

So began many hours of talk spread over five years.

She is Portuguese. This fact explains many attitudes but has generated some problems. Being Portuguese is a thoroughly physical identification: a breathing, such as the Greeks meant when they spoke of the living breath as *pneuma*—a warmth, a vitality, a sparkle. This identifies her, and she accepts this identification happily. But on the other hand, her mind rejected the old Portugal vigorously, and that was highly divisive. Essentially she rejected the ignorance she considered a stigma in Portuguese history. She called it "demi-ignorance": There is nothing wrong in not knowing, but there is something very wrong if you know a little and act as if you know it all. For a nation such an attitude is suicidal. It is, however, given twentieth-century history, easily understood.

A dictatorship, or any totalitarian government, has the effect of freezing a moment in history, crystallizing the institutions, beliefs, attitudes of those in power. The consequences both for a country and for individuals can be lethal. Salazar was dictator for nearly fifty years, and Anna equates the demi-ignorance of Portugal with his regime. It was a dreadful regime, not only in the obvious sense of putting people in prison and torturing them but because it destroyed two or three generations. Pervasive and subtle, it continued so long that complacency rippled outward from the dictator to the people. They proceeded to turn their backs on the rest of the world and came to believe that in their attitudes and beliefs they alone were right and had nothing to learn from anyone else. To go to Portugal before the revolution was to go backward into time.

"When was your mother born?" she once asked me.

"Eighteen ninety-eight," I replied.

"Then remember," she said, "that she who was born before this century and I who was born forty years into this century had the same sort of education—the same kind of upbringing."

For she was the only child, and at that time an only female child was expected to be a loving daughter, get married early, and produce grandchildren. She was not expected to become a medical doctor, reject her country, leave home, and stay away. She came from a financially comfortable background that reflected many of the accepted attitudes of Portuguese society. Even so, comfort was never equated with idleness, and she was expected to excel at anything she was taught. Nor were her parents entirely typical. They were anti-Catholic and egalitarian, and there was no question of her going to private school with the rich girls. She went to public school with the poor ones.

One experience stands out from these early years. She was six or seven years old, and the whole class was struggling with the problem of multiplying by zero. The teacher, Donna Fausta, a lovely old woman, set what appeared to be an enormous challenge. "Try to find out what happens when you multiply things by zero." For most children, the very idea is silly. "Nothing" happens, but why should the answer be "nothing"? You have one thing in your hand, and you are multiplying it by nothing and still you have the thing in your hand. How can zero and "one thing" be made to fit? Anna was fascinated. She remembers sitting there, multiplying away like mad, and what quickly came across was the notion of an abstraction. She describes this as a fantastic moment. So now the problem was: How could one multiply a "thing" by an abstraction? She became entranced by the idea of a pure, abstract, transparent thought—and has remained entranced.

Another enthusiasm of her early years has also endured: She loved to be alone. If everyone went out she was delighted. It wasn't that there was anything in particular that she wanted to do; she might play, or read, or practice the piano. Solitude was the essential point. She remembers this as provoking a physical experience, one that actually reflects the quality of science. It is the sensation one has when entering a library, a feeling both of space and physical well-being. This love—this need—of empty space may have grown from many hours spent on the shore. There was a long beach which, in one direction, was protected by many high rocks but in the other direction was endless empty space. And that was not only where she went but where she felt secure. Mountains are confining, even terrifying; walls must not be so high that they cannot be seen over.

Portugal may have been fossilized in certain respects, but the secondary school was old-fashioned in the very best sense—as indeed was my mother's. It was in an old palace, and might have had water coming in through the ceilings and paint flaking off the walls, but Anna remembers it as a *real* school, with superb women teachers in biology, physics, and mathematics. She was still passionately devoted to mathematics, but while the challenge of zero was still there, geometry was even more seductive. She tried to get ahead of the book and invent the theorems before they came up; she soaked herself in the early Greek philosophers and mathematicians—"That lot, on those Greek islands, who invented the triangles and spent all those years . . . just thinking!"

The teachers at school put her in touch with ideas whose echoes she still hears. One, which she found in a biography of Pierre and Marie Curie, was encapsulated in Pierre Curie's saying: "It is necessary to make life a dream and of that dream a reality." She both loved and understood the aphorism of Jean Jacques Rousseau: "Hypotheses are the revelation of genius."

When finally, just short of her seventeenth birthday, she decided to become a doctor she went to the medical school of Lisbon University. Judged by any standards, it had a very orthodox—even old-fashioned—syllabus, with two years of anatomy and physiology followed by clinical work. But the courses were good, and she feels that in many ways the system was far better than the one which now prevails in America, where the students are forced through batteries of tests and exams every term, "just as if they are small kids." She remembers the students as having great stretches of time in which they were encouraged to range intellectually. Six years later she was qualified, but by then she had turned away from medicine toward scientific research.

The turning point came while she was doing her clerkship in ophthalmology. There was an institute of ophthalmology in Lisbon just for the poor, and there the medical students learned their ophthalmology by standing alongside a doctor and watching a line of people approach. They stood still, then came forward as pleading pairs of eyes. You saw the eyes, pronounced, and dismissed them; saw the eyes, pronounced, and dismissed them. In particular, Anna remembers one crucial couple, a fisherman from Nazaré and his wife, who was a dressmaker. The fisherman worked on the docks, on the crane, and had gotten a piece of metal in his eye. The piece was microscopically small, but still he felt pain, and it had been in his eye for three months by the time he and his wife had saved enough money to make the difficult trip down to Lisbon. His wife had terrible headaches too, without understanding why. The lenses in her only pair of glasses were broken down the middle and she could not afford new ones. They stood in the line, and as the two pairs of eyes moved forward, Anna knew she was not going to practice medicine in Portugal. Perhaps she was not going to do medicine anywhere.

She recognized two quite separate reasons. Here were the fundamental people of her country, expressing its very soil, to whom it was impossible to respond except with compassion and pain. She realized that she couldn't take it. Thus her decision to leave medicine was a decision not to be destroyed. The second reason came from her sudden realization of the core problems in Portugal. They were not really medical. They were political, economic, and social. So she had to make a decision: Did she want to be a revolutionary and go to jail, and perhaps do *nothing* for these people? Or did she want to get out, establish herself the long way around, if necessary, and perhaps help them in some more effective way? Yet she always hoped, and hopes to this day, to prove that the Portuguese *can* overcome their ignorance and self-satisfaction, and if she could achieve anything as a Portuguese, and anywhere in the world become a good scientist, then perhaps she would be in a position to help effectively. This, she well realizes, can be regarded as pure rationalization: "Rationalizing my selfishness. It sounds terribly generous and high-minded, but still I went on to do what I wanted to do and what I liked." She would become a scientist. The decision, she recalls, was utter misery to make and very difficult to carry out, for many painful things followed as a consequence.

In her last year of medical school she had to write a thesis, and to this end she worked in the Lisbon Sanatorium with a pathologist, looking for evidence of precancerous conditions in chronic lung diseases. It was her first introduction to scientific investigation. There were questionnaires to be made up and sent out to patients, dealing with such matters as whether or not they smoked, and how, when, and how much. She remembers getting back the most irrelevant, irreverent, and rude replies. But the project gave her the flavor of scientific research and, more importantly, since she had to examine the tissues of the lung most carefully, gave her the chance to realize that she thoroughly enjoyed work with the microscope.

She might well have stayed in Lisbon, but that same year a group of young people decided that they wanted to build a new research institute, different from anything that had ever existed in Portugal. They invited Anna to join, and she, of course, would join anything that was going to be different from what had gone before. The man who organized the enterprise asked her if she would be prepared to study immunology, to take postdoctoral training abroad and then return to help build the new world of science and medicine. She believed in the dream absolutely, although she was, she now admits, completely naïve, politically inept, and ignorant as well. She was given a choice of Australia, Sweden, or Great Britain. Sweden seemed too cold, Australia too far; she chose Britain.

She was uncertain about the move because she felt she should have more clinical experience, but everything went forward rapidly. A scholarship from the Calouste Gulbenkian Foundation gave financial help. The Imperial Cancer Research Foundation provided a scientific base in its Mill Hill Laboratories in London. The director of the "dream" institute-to-be came back to Lisbon from visiting London, full of stories of lovely red-brick houses and laboratories among trees.

Anna defended her thesis in October 1963 and was awarded her M.D. She had never left Portugal before, had never traveled by plane, but on February 29, 1964, age twenty-two, she finally departed, carrying two books which were presents from her mother: a textbook of pathology and the classic textbook on immunology written by Drs. John H. Humphrey and Robert G. White. Like it or not, she was on her way.

She was to join the laboratory of a Professor Vera. She didn't really want to go. And, as a matter of fact, no one in London wanted an unknown, untried, visiting "underdeveloped" investigator either.

CHAPTER 2

TRADITIONS

Science, at bottom, is really anti-intellectual.
It always distrusts pure reason and demands the
production of the objective fact.

H. L. MENCKEN, Notebooks Minority Report, I

When we met, that February evening in New York, Anna's sabbatical still had two and a half months to run, and we started work at once, recording the background so that I could grasp her current scientific situation and its outstanding problems. Very soon I came to know the world of the laboratory at night, for it is in those early morning hours poised uncertainly between two days that Anna works best. Then the concentration can be absolute, thoughts come without effort, intuition takes hold, and past experience leads. The silence and the isolation that concentrate the darkness epitomize the internal life of the scientist. In essence, the world of the mind is as tranquil as an empty room, and Anna was often to speak of the way in which the discovery of a new fact related to the blackness of the unknown "like looking through the keyhole into a dark room, or gently opening a door and there is just one small pinpoint of light inside." It holds promise but no more, for what unexpected thing might be lurking in the corner?

My first protesting introduction to this world came one night in February 1975, not long after I first met Anna, when at 11:30 P.M. I made my way to the entrance to the laboratories on the East Side of midtown Manhattan. The doors were firmly closed. A group of malevolent deities were picking up the entire Atlantic Ocean and happily sloshing it over the city, and I waited outside for a while, getting wetter and sleepier and more reluctant. Finally I turned back on my way to bed. But as I walked past the entrance to the hospital a white-coated figure dashed out into the rain, grabbed me, and led me inside and down into the bowels of the earth. The tunnel joining the hospital and laboratories was dark, warm, and comforting. I was taken up in the elevator for six floors and led along the corridor; then Anna opened a door.

The quiet pressed down from the low ceilings of a room filled with instruments. The silence was metallic. One light alone was on, in the laminar-flow cabinet, a hooded work area with a powerful fan which ensures that only clean and sterile air passes over the bench in front of the scientist.

Anna put on a pair of transparent, thin, whitish gloves, saying as she did so, "I hate this job because I daren't think of anything else while I'm doing it."

"Isn't this a routine job for technicians?" I asked.

"Yes, it is. But it's terribly critical, and when one mistake can ruin the experiment, I try to do it myself."

She picked up a pipette, sat down at the bench, and went straight on with the work. Against the single light she was a silhouette, head and back tense with concentration. Rows upon rows of small tubes stood in a rack in front of her. Flat dishes, twelve inches by eight, each with a hundred or so small depressions on its surface, were on the bench too. The hands went up, the pipette was placed in one of the depressions. Over and over and over again; in and out, in and out, biting her lips as she worked.

I took the hint and moved quietly around the room until I found a stool and a place at a bench where I could sit. The microscopes were shrouded like the guns of a ship in harbor, but the wind and the rain still pounded at the windows. Then I became aware of other noises, and as I turned, my head moved through several spheres of sound. The fan of the hood was the loudest noise of all, whooshing, pervasive and regular. But other noises made themselves known by their quiet persistent regularity. The *plonk-plonk* of a water bath, constantly stirring automatically. The *tick-tick* of an electric cell counter. There are, I was to find, musical rhythms in all places where counting goes on, and these days most laboratories working with cells end up by counting them.

Tock, tock, tock, the cells add up. Then when the set time is up there is another little click, followed by a cascade of sounds as the counter is set back to zero. *Cling-clang,* the fridge switched on and, grumbling cold, joined in the chorus. And one little red light blinked as the spectrophotometer continued to measure light absorption in silence. The machines are set to continue science when the scientists are not there.

After half an hour, nothing had changed. The job could not have been simpler: just mixing fluids, adding something that makes cells divide to cells whose dividing activity—mitogenic activity—you wish to measure. The "something" you add contains *mitogens*, the substances which precipitate the dividing. But the sequence and the concentrations and the amounts of the mitogens were both variable and crucial, and I could understand that the pattern of thought and movement of hand had better not be broken if the procedure was to be accurate.

But after three quarters of an hour, I was asking myself, How long, O Lord, will this go on? How many minutes of how many years would I spend in watching such routine, prosaic, and boring tasks? Surely when you have seen one pipette going in and out, you've seen them all? Yet clearly there was a deep point both in the task itself and in the repetition of it, and part of my job was to find that point. The hands performing this routine job of deceptive simplicity were, no less than eyes, ears, microscopes, and scalpels, tools in the service of a person with ideas, who had, for the duration of the task, to turn her head into a vacuum.

The time passed. What was she doing, I asked myself, and *why* was she doing what she was doing? The end point of the task and the repetition was, of course, a "scientific result"—a phrase that encompasses innumerable activities whose real point is understanding the world in which we live. "I'd like to be able to carve my results on a Henry Moore statue," she was to say later, meaning not only that the results should be good enough and enduring enough to justify such an act but that statue and inscription together would bear witness to the same creativity at work in carving a sculpture and constructing a scientific theory.

In this case, the theory was the one she had spoken of before and had to do with those same small cells, the lymphocytes, complicated in structure and exquisite in function. They are a crucial part of the lymphoid system that maintains "the integrity of the body," which phrase—it is the title of a book on the subject by the distinguished Australian immunologist and Nobel Laureate Macfarlane Burnet—sums up the nature and purpose of the immune system. Because of this system, we hold ourselves together; without it, we would fall apart. And the essential property in maintaining that integrity is the capacity to recognize "self." The prime function of the immune system is not to reject foreign transplants, as one might think in this sophisticated age of skin grafts and heart surgery, but to distinguish a cell that does not "belong" from one that does and to reject it, thus allowing the cells of each individual to stick together and the organism to persist as a unique, single entity. This basic function is possessed by all vertebrates (and also had precursor functions in simpler forms) because all animals need to repel invaders, resist infection, and destroy aberrant cancer cells.

"Invade," "repel," "attack," "destroy"—in the peace and tranquillity of the laboratory night the military analogy seemed obscenely incongruous. But it is the most effective analogy nevertheless, for there is a constant war between the body and those subversive agents that attack and sometimes destroy it. Living warriors—bacteria, viruses, and parasites—are matched against other living warriors, the cells of the immune system, commonly known as the white blood corpuscles.

Basically there are three kinds of white blood cells. The yellow pus of a skin infection indicates the presence of *polymorphs*, the light infantry of the body's battalions. Moving fast, they rush to the site of a bacterial invasion to spray a fusillade of lethal chemicals— lethal, that is, to bacteria. Polymorphs live but a short time and can themselves be poisoned or deceived.

The next group are the *macrophages* (*macro* = "big," *phage* = "eater"). Large and sluggish, like the heavy infantry, they arrive on the scene later. They are found all over the body, in the tissues, in the blood, and especially in the spleen and lymph nodes, those filtering/screening stations of the organism. Macrophages live much longer than polymorphs and can also reproduce themselves. But they do not shoot; they devour.

In their actions both polymorphs and macrophages are quite effective, but they are not sophisticated. They attack almost anything and thus give us a general protection, what biologists call "nonspecific immunity." But invaders are more subtle than that, and just as a sophisticated, precisely targeted missile calls for a sophisticated and precise defense, so too does the body require a sophisticated response to the subtleties of such attacks as smallpox and diphtheria.

This response is provided by the cells that Anna had in liquid suspension in the test tubes on the bench: the *lymphocytes,* a population of remarkable cells which are crucially important, for they confer the property of specific immunity.

Some lymphocytes are round and have knobs; some are round with projecting fingerlike structures. They must make their way through the interstices of the body; insinuate themselves into the fluids that bathe all the tissues; take station in vital organs like the liver, the spleen, and the intestine; emerge from the blood vessels in order to slide back into their own circulatory system—the lymphoid. To do this they must change their shape from a sphere with knobs to a spook on Halloween, flat like a planarian worm, emerging out of the caves of the blood-vessel wall. I saw a photograph of one taken as it entered a minute vein, rising up stealthily like the head of a cobra, definitely not something to meet on a dark night while at war.

At any one time, each human being's body contains about two thousand billion lymphocytes. Some live only a very short time. These lymphocytes are constantly being replaced, in order to keep the immunological balance, at the estimated rate of one million cells every second. Born in the same place, the marrow of our bones, and from the same parents, the stem cells, young lymphocytes initially all look alike. But some—the T-lymphocytes—move off to the "finishing school" of the thymus gland to acquire the final sophisticated touches that enable them both to distinguish those cells that belong to the body from those that do not and to attack specific foreigners like tumors or transplants. Without a thymus gland you are in serious trouble, immunologically speaking, for it is in this gland that the adolescent T-lymphocytes, or T-cells, as they are called, develop those surface receptors that allow them to recognize foreign cells or foreign substances, known as *antigens.*

Eventually the squads of T-cells contain both reconnaissance scouts and challenging sentries, as well as killer commandos and commanding officers. For it is from one set of T-lymphocytes that instructions are sent to the second great class of lymphocytes.

It is not yet confirmed just where this group—the B-lymphocytes—mature. But wherever they grow up, by the time they are through they are able to "neutralize." Exposure to the smallpox antigen, or to one of diphtheria, for example, somehow leaves a mark on our immunological memory. Receiving instructions from the T-cells, the Bs will secrete antibodies at the rate of some two thousand molecules per second. At the same time, the bone-marrow cells are "instructed" to form a new population containing the memory of the encounter with the foreign substance. This population remains passive and quiescent until the next skirmish with smallpox or diphtheria. Then, with a burst, the population that secretes the counteracting antibodies explodes in quantity, and the system is flooded with the antidote.

Finally there are the *lymph nodes*, glands in the neck, the groin, and the gut. The middle of the spleen is actually a lymph node. These nodes are the filtering stations, places where blood and lymph systems converge and all sorts of biochemicals are filtered out. They were discovered to have this function when traces of ink were found in the lymph nodes during an autopsy on a dead, tattooed sailor! When in earlier years Grandma said, "The child has swollen glands," she was merely pointing to evidence of the terminal stages of an immunological battle, a lymph node packed with more lymphocytes than normal.

The question "Why did the lymphoid system evolve at all?" is one that has tantalized biologists for a long time and is not yet fully answered. In fact, our present accepted knowledge is not twenty years old, although the Egyptians knew about glands well over two thousand years ago and therefore about the lymph nodes. Swollen glands in the neck are swollen lymph nodes, and the Egyptians, good observers that they were, and good surgeons too, would notice a suppurating gland and open the neck to see the scrofulous white material hanging inside. Then they chopped the gland out and packed it in herbs. The Greeks knew about the glands too. Hippocrates wrote a book about them in which he classified the thymus gland, the lymph glands, the mammary glands, and the brain "gland," according to their looks, their consistency and their whiteness. He included the brain, so Anna told me, because his logic went like this: Everything that is white is a gland; the position the glands occupy in the body is related to the absorption of the liquid "humors"; where the body is dry, there are no humors and there is no hair there either, so the glands, all consisting of softish white material, are always associated with the presence of hair. The nodes in the groin are near the pubic hair: those in the neck, near the beard. The white stuff in the body where the longest hair is to be found—in other words, the brain—must therefore be the biggest absorbing gland of all and take up most of the humors!

Some two thousand years later, in the middle of the nineteenth century, the German pathologist Virchow argued that the lymph nodes and the spleen were not just filters but centers where the white blood corpuscles were made. The corpuscles were, he believed, unimportant short-lived cells, with no obvious function, just simply taking a brief journey through the body. A small gland at the base of the neck, the *thymus*, seemed equally unimportant. The knowledge of *its* function is even younger in scientific time, not yet thirty years old. Before this, medical fashion, not scientific understanding, had sometimes dictated that in very young children with persistent colds and swollen glands in the neck, the thymus could be usefully irradiated away. The consequences were disastrous.

The true role of the lymphocytes in the body could not even begin to be elucidated until the late 1950s, when Dr. J. L. Gowans, now director of Britain's Medical Research Council, initiated studies in Professor Howard W. Florey's laboratory in Oxford. By using a technique to drain the thoracic duct, a main channel of the lymphoid system, he was able to find small (five hundredths of a millimeter) lymphocytes in the fluid he extracted. Thanks to a new, postwar, post-Manhattan Project technique of labeling the cells with radioactive isotopes, these cells could be placed back in the animal and their destinies traced. Thus Gowans showed that a second circulation exists in the body in addition to the bloodstream. Small lymphocytes move out of the bloodstream into the lymph, via the postcapillary venule, a minute vein in the lymph nodes, to wander over, around, and through the tissues of the body, looking for trouble. Then they re-enter the bloodstream via the lymphatic vessels which open into the large veins.

It took just ten years for scientists to dismantle the old classic view of the lymphoid system, the ten years between 1956 and 1966. They have not yet completed their reconstruction. But so crucial was that decade, and so intense and remarkable the work, that it took on the characteristics of a blowout in an oil well. The constraining cap of traditional thinking shot off with great force, and the pent-up pressures of interested scientists were released into a burst of furious activity. Anna has written about it in these terms:

In all probability, that decade will be described by historians of science as an epidemic of understanding of lymphocyte function, rather like a creative and intellectually procreative form of encephalitis. An infectious interest in immunological function swept across the world from Australia to England, to Connecticut and Minnesota in the United States; across species from the chicken . . . to the mouse . . . to the rat; across the board of scientific activity from experimental work . . . to theory. . . .

If, in February 1975, Anna was doing an experiment with lymphocytes rather than, say, with nerve cells, in a cancer laboratory in New York City, it is because in 1964, as a young medical doctor, she had left her native country and gone to England and there succumbed to the prevailing epidemic.

CHAPTER 3

NEW BEGINNINGS

"The Regiment is going to advance.
Send reinforcements." MESSAGE SENT

"The Regiment is going to a dance.
Send three and fourpence." MESSAGE RECEIVED

She arrived in London toward the end of that remarkable decade 1956–66, when the tide of research on the lymphoid system was in full flood. The landmark discovery of the indispensable role played by the thymus gland in the regulation of the immune system was linked to a crucial observation: Removing the thymus from adult animals had no effect whatever. Removing it immediately after birth, however, resulted in a deficient immune response, which was shown when the animals thus treated then accepted foreign skin grafts. Ultimately the animal died—from gross immune deficiency, we know now, for half its immune system was missing.

During the years 1960 to 1964, similar evidence and similar discoveries about the thymus were being reported in a variety of publications by scientists who had been working with different animals. The advance in immunological understanding took place on a very broad front, and many factors, from luck to habit, influenced the order in which such reports were published and thus the priority in claiming credit. Scientific publications have different lead times and different lag times. A letter can sometimes be published in *Nature* very quickly, as happened when Watson and Crick reported the structure of DNA. On the other hand, the *Proceedings of the Royal Society of London* and other similar journals can sometimes take upward of two years to print an article they have accepted. At any rate, in a three- to four-year span, many studies of the thymus function were published: on the rabbit, mouse, and chicken (Dr. Robert A. Good in the United States); on the rat (Dr. Byron Waksman in the United States); on the mouse (Dr. Jacques Miller, working in England, and Dr. Delphine Parrott, also in England). Everyone was looking at the thymus, and all were correctly concluding that it was crucial to the development of immunity and the production of lymphocytes. But they were not certain whether there was one population of lymphocytes, or two, or even more; and if there was more than one population, was the thymus gland responsible for them all?

With questions like these in mind, the head of the laboratory Anna was to join had made a collection of microscope slides. The slides were the sectioned organs of mice, some of which had had their thymus glands removed—the thymectomized ones—and some of which were normal. A few of the thymectomized mice had died "naturally," but most animals—experiments and controls—had been deliberately sacrificed at various stages, from the newborn right through to the adult. The slides were arranged in a chronological order which roughly corresponded to the time the animals had had their thymuses removed, the lengths of time they had lived, and the stage at which they had been killed. Thinking that the sectioned organs might indeed reveal something—though she would not guess what—Dr. Vera had sent them to two of the finest pathologists in London. Each separately assured her that there was nothing of significance to be seen apart from some hyperactivity of the plasma cells, those cells whose function is to produce antibodies. This apart, it certainly wasn't worth anyone's spending much time with the material.

In the matter of the visiting scientist, Dr. Vera had no choice. One day the director of the Imperial Cancer Research Fund (I.C.R.F.) informed her that a Portuguese woman was coming to the institute to train with them. She was somewhere in her late thirties, had a lot of scientific experience that included a great deal of animal pathology and cancer research, and, of course, was fluent in English. Dr. Vera had also met one of the directors of the new institute in Portugal, to which Anna would eventually return when her training was complete, and from him, too, she was given to understand that the visitor was terribly bright, terribly experienced, in fact, "terribly everything."

Everybody, it turned out, had been thoroughly misled, including Anna herself. Her prospective boss in Lisbon actually wished

her to train as a virologist and had said this to the director of
I.C.R.F. It was not until many years later that Anna learned that his
real intention had been to have her learn those aspects of virology
and immunology which would be relevant to his own future work.
Therefore, all the virologists at Mill Hill had been asked if they
would train the visitor, and the answer had been no.

Anna arrived on a weekend, just as Dr. Vera was going off to a
scientific meeting in Holland. She phoned Anna at the Gulbenkian
Foundation, welcomed her warmly but briefly, and added, "Don't
come to the lab until we are back." But Anna *did* go to the lab, and
when Dr. Vera returned from Holland, there she was, already installed. Dr. Vera's first view of the new arrival was of a short,
round-faced, dumpy Portuguese wearing a lab coat much too large
for her, which made her look even dumpier and shorter. She was far
too young and wore an expression of great amiability combined
with what looked like total incomprehension, even stupidity.

It was quite clear to Dr. Vera that the new arrival was not going
to be welcome in the laboratory at that particular time. So the very
first morning she kept Anna alongside her while she did some routine operational techniques on mice—taking out their thymuses—
and, while she worked, tried to find out just what the foreigner
knew, or could do, or could even *say*. She swears now that Anna had
never *seen* a mouse in her life, dilapidated Portuguese school building notwithstanding. Indeed, her only experience in pathology had
been strictly limited to human material.

Thus the first move was to place her where she could be super-
vised but not put off by an unwelcoming atmosphere; the second
was to try to think up something she could *do*. She was installed in a
building named the Hut, which had a pleasant view of grazing
horses, and she was lent a microscope until such time as one would
be purchased for her. Finally, Dr. Vera handed her the whole slide
collection of mouse sections and told her to sit quietly and look at
them, "just for fun." Then she was to report on what she could see.
Anna is short and the laboratory stool was low. The bench, on the
other hand, was high. If she put the microscope on the bench and
sat on the stool, she couldn't see down the eyepiece. So she opened a
drawer halfway down the bench, popped the microscope into the
drawer, and looked down the microscope from a more convenient
height. She reconstructs Dr. Vera's actions in these terms:

She must have thought: Here is this nit from Portugal, and the only thing she can do is look down a microscope. So we might as well give her something to look at, and so as not to frighten her, we might as well tell her that it doesn't matter if she doesn't find anything. That is why she said, "Look just for fun." This is a crucial joke now. Everything I have ever done since then has been just for fun. That was really very good.

From time to time, Dr. Vera would come in and see how she was getting on. Anna didn't say very much; in fact she *couldn't* say very much, because at that time her English was quite limited. Whenever she wished to communicate with Dr. Vera, she would turn around, pin some paper on the wall, and draw what she wanted to say. Thus the conversation, though friendly, was somewhat restricted, limited to the drawings on the wall and to the cells down the microscope. And though Dr. Vera made it clear that she was available if help was needed, none was ever asked for, except once when Anna needed a translator to help negotiate the rent for a bed-sitting room.

The most prosaic of circumstances and the most unlikely of situations can nevertheless be the source of intensely exhilarating creative experiences. Anna consistently refers to those that derive from creativity or discovery as imparting the conviction of "belonging." Even though what is discovered has always been there, the scientist seeing it for the first time has the sense of calling it into being, and thus of belonging not only to the moment of discovery but also to the fact discovered and even to the scientists who will study it next. This is why she maintains that making a discovery is not so very different from being in love or making love. You can be in an enormous desert in respect to your personality and emotions; then you see someone, love them, and you belong. In the same way one is in an enormous desert of space and nature; then you see something new, understand it, and again you belong.

The first moment of illuminating observation may have been intensely exhilarating, but from then on it was all struggle, lasting nearly a year. She had arrived in London on February 29, and by the end of March she knew that she really had discovered something. Day after day she had sat in her little corner, looking down the microscope, but scientists don't think too highly of other people who only look down microscopes. She didn't do any experiments; she hadn't really done any experiments in her life. In any case, she thought that sitting in her little corner, being "as quiet as the mice," was what she was supposed to do, and that is what she did. Periodically she tried to tell Dr. Vera just what it was she was seeing, but at first Dr. Vera didn't understand and, when she finally did, she didn't believe it.

Four months after Anna arrived, on June 30, 1964, Dr. Vera went to Anna's room and said that since she didn't seem to be doing very much, or seem to be very interested in the work, "perhaps she would like to go home." This was a very English way of putting it. The Gulbenkian Fellowship still had months to run.

Dr. Vera now says she was being grossly unfair, but to be fair to her in turn, one must remember that she was being pressured to get rid of the "nuisance" who was occupying space and tying up a microscope. And she did think Anna's progress was a little slow. She had expected to see some light dawning—even though she believed there was nothing significant on the slides—or hear some intelligent comment about something. Anna was absolutely furious but altogether silent. "I was bloody angry. I thought she was incredibly unfair, because I knew by then that I had seen something new." But after four months she had already acquired some of the mannerisms and manners of the English, and she knew that in public one pretended not to have emotions.

But what *had* she seen that she thought was really fundamental? In simple terms, this: She had noticed a fact that others had missed—that there were great empty spaces in the center of the spleens and lymph nodes of those mice that had lost their thymus glands.

The spleen in any animal consists of two parts: an outer red pulp and an inner white pulp, which contains a lymph node called the *Malpighian body*. Lymph nodes anywhere are all structurally similar, with an outside layer and a central core packed with lymphocytes, but in animals that have lost their thymus there is, within twenty-two to fifty days, a depletion of the lymphocytes from the central core of the spleen and the lymph nodes. This depletion shows up on microscope sections of these organs as a blank white space in the white pulp.

Why this empty space? Anna had worked out the reason: There

is an empty space because the T-lymphocytes, those lymphocytes that need the thymus gland to mature, are missing, and there is nothing to fill the space until, after some fifty days, other cell types gradually drift into the blank areas. *But no lymphocytes do.* Thus she realized not only that there are two distinct populations of lymphocytes but also that these populations remain distinct. In the adult animal the T-lymphocytes always occupy the central or "thymic-dependent" area of the spleen and the lymph nodes, while the second population occupies the "thymic-independent" area surrounding the central core. Why each lymphocyte population occupies its own separate and specific zone is a question she is still trying to answer.

Dr. Robert A. Good, one of the pioneers in thymus work, assesses Anna's discovery in these terms:

> Her work was, I think, *the* crucial work in that regard. I was very much impressed with what she had done. You can talk about the fact that the field was ripe, and it was. There was no question about that, because Dr. Turk was coming at it from another point of view also, and maybe all of us would have come to the same place. But before we ever got around to the mouse, she came up with this entire definition of separate lymphocyte populations: that in the adult organs there were thymic-dependent areas. And she discovered this on the basis of her studies, and her careful analysis of the histopathology of the mouse lymph nodes in the spleen. All sorts of people had been looking at the material and had not seen—simply not seen—the sharpness of the disassociation. As a novice coming into the field she just cracked it, just like that. It was her work that gave us the picture.

Why did a novice coming into the field manage to crack the problem whereas other, far more experienced people had not?*

Dr. Vera says that, first, it was because the novice had a "clean" mind, a mind not cluttered with any preconceptions, either of what lymphoid tissue ought to look like or with previous knowledge of the material under the microscope. She was not only able and willing to look at the material under the microscope with unprejudiced eyes; she was bound to do so. But secondly, she was tantalized and wanted to find the reason why some of the thymectomized mice

---

*Dr. Byron Waksman had also seen the phenomenon of localized lymphocyte depletion in the rat and guessed its significance. But he did not tie in this observation to the circulation of lymphocytes or do any experiments on this point.

had died "naturally." (No immune response!) Other people had blithely assumed that these mice had died because they just suddenly took sick. But Anna did not make that assumption, and so she was painstaking and willing to look and look and look again; to go back and forward with the slides; to look systematically at the sequences of age, time of thymectomy, and death; to be utterly patient; to spend literally hours at the microscope rather than pick up one or two slides at random and give a snap judgment. That is what the two pathologists had done and what most pathologists regularly do at post-mortems; they pick up two or three sections of material, look at them, and say, "The patient died of X."

Anna may remember her willingness to spend hours at the bench but does not now consciously remember all her questions. She does remember feeling that Dr. Vera must surely have had some point in giving her the slides, and therefore they had to be studied carefully. When I asked her why she thought others had not seen—let alone appreciated—the disassociation referred to by Dr. Good, she replied, "Somehow they seem not to have been able to for various reasons. For one reason they tended to look only at the slides of the later stages, and by then it was too late—both in the sequence of sections and the age of the animal. The crucial changes had occurred earlier, and there was nothing after that to follow up. The fact that Dr. Vera had mice of all sorts and of all ages enabled me to start at the very earliest stages and pick up a gradual progression in the story of the missing lymphocytes. So I consistently asked for the various age groups until I could trace the development through time. Also I did have a very good training in pathology, and I *was* a good observer."

She may have been furiously angry when at the end of June Dr. Vera suggested that she might like to go home, but she says now that it was the best thing that ever happened to the "nice little Portuguese girl." Eventually it did filter through to Dr. Vera that her visitor was very angry indeed, and finally they decided to leave things as they were while Dr. Vera went on holiday. By the time she came back Anna had recovered her poise and was able to thank Dr. Vera for "the shake," which so far as she was concerned had done two things: First, it had shown her that she could, and must, plan experiments and do them, and second, that she really did have genuine scientific questions—that she hadn't just been sitting there like a dummy. This realization in itself proved a source of confidence. The episode also showed that she had to look after herself, and, if she intended to carve out a scientific career, she had better start right there and then.

"So I became much less soppy and much more defiant. Also I became an experimentalist. Before this I was an observer, a creative observer possibly. But then I went to work with two women who were first-class experimentalists. So I became a scientist."

As for Anna's observation of the mouse sections, of course, Dr. Vera still couldn't believe it. She didn't in fact believe it until almost a year later when an experiment proved it. She herself had anticipated some depletion of lymphocytes in the thymectomized animals, and was not surprised about that, and she herself had earlier observed the areas of depletion even to the extent of seeing that parts of the lymph nodes were apparently empty, without a lymphocyte in sight. What did surprise her to the point of incredulity was the specificity, the selectivity of the depletion Anna pointed out—not only that there were empty areas but that they were so clearly defined that later scientists were able to say, "Normally the T-lymphocytes would be right here, and the others there." But basically Dr. Vera was an experimentalist and thus extremely skeptical of all pathologists, and because she respected those people who did things with their hands at the bench she would not draw a conclusion on observation alone. The pathologists were "the dead meat people," who could only explain away somebody else's genuine scientific questions.

But there was one observation which was really sophisticated, subtle enough to make her stick with Anna's claim and not reject her finding out of hand. Lymphocyte destruction naturally occurs in all animals that have had a viral or other infection. In such cases one *expects* to see empty areas in the lymph nodes. But Anna had been sharp enough to spot similar empty areas in the tissues of *healthy* thymectomized mice, where gross cell destruction would not be expected to have occurred. This was a puzzling clue—to something, an impressive observation that cried out for explanation, a fact suggesting that Anna was on to something important.

But it is all very well being in a "desert" and seeing something you think no one else has seen. It is all very well to take an existing situation and try to refract a different light through it, to make another picture for other scientists to see. The problem is to persuade the world—or the community of your colleagues—to see things the way you now see them—i.e., differently. Albert Szent-Gyorgyi was once in a group with other Nobel Prize winners and was being plagued by the press with questions like "What did you feel when you made your discovery?" And Szent-Gyorgyi said, "I felt bloody angry. Because there it was, *the thing,* and I was the first one to see it. But then I had to go and tell all these other people." You can have your moment of illumination, and it can be a mystical experience, an exhilarating experience, but unless it comes alive for other people it remains just what it was before—a stone in an empty desert.

It took a full year to persuade Dr. Vera of the fact of Anna's finding, and then of the significance of the finding, and then of the truth of the finding. But it did become a little more likely when a few months later a paper was published by some famous scientists, and in the photographs illustrating their paper the depleted areas were clearly visible. The authors themselves neither saw that fact nor its implication, but Dr. Vera could see that their photographs matched what Anna said she saw on the slides and had drawn in her diagrams. Nonetheless Dr. Vera's continuing cautious reluctance reflected the natural and proper skepticism of the scientist and the fact that, as an experimentalist, she really was unable to perceive the fact *without* the experiment. One may have a good microscope eye, one may see something that is genuinely there and make a valid discovery, but in the absence of a demonstrable specific function, established by experiment, immunologists are not going to be impressed. For them a thing exists only when the function has been demonstrated.

The months between June and September 1964 were the first months in the making of a scientist. "The design of my first experiments was awful," Anna recalls. "Dr. Vera called them diabolical. This is always so with scientific beginnings." By then Dr. Vera had no intention of making Anna a highly skilled technician who could learn a couple of techniques and then go back to Portugal and practice them for the rest of her life. She may have been skeptical of her findings, and would never accept them without careful proof, but she had sensed Anna's potential and was going to make her a scientist. And that is precisely what she did. Anna says, "I learned so much from that woman—it is unbelievable. It is like sculpture. The scientist I am today has been carved down from the person I was. Cut, cut, cut; all the extraneous mess cut right away."

Early that autumn Anna had brought Dr. Vera her first carefully drawn diagrams of the mouse slides showing the clearly defined empty areas on the lymph nodes and the spleen. She pointed to the zone in the middle of the spleen and said, "This is the thymus-dependent area," and, pointing to the area surrounding the inner zone, said, "This is the thymus-independent area." When Dr. Vera protested that she couldn't call it a "thymus-dependent" area because she didn't know whether it had been inhabited by T-lymphocytes, Anna had replied that she could *see* it had been. But she was informed that she still couldn't say so until it had been proved—until she had shown, in fact, that if there are holes in tissues because of missing cells, you can fill the holes in by injecting the right cells.

All that Dr. Vera was prepared to admit at that time was the *possibility* of a prediction. If they labeled the T-lymphocytes with radioactive isotopes and injected them, and if Anna was right, the lymphocytes should turn up in those thymus-dependent areas where Anna predicted they would. Recalling this time, Anna remembers being horribly nervous. If the cells behaved as she believed they would, "I'd be in luck." Indeed she was. The experiment was done; radioactively labeled thymus-lymphocytes were injected into some mice which had had their thymus glands removed and into the controls which had not. By autoradiography—that is, by sectioning the organs and taking photographs of the radioactively labeled cells—their route through the arteries and the capillaries could be followed and their arrival into precise areas in the spleen and lymph nodes established. When Dr. Vera saw the clinching evidence of the autoradiographs, showing the cells in their predicted places, she said, "I think I begin to understand." Anna was moved to tears: "It was an unforgettable moment."

By this time she had been moved from her isolated corner with its view of the horses to an upstairs laboratory. The unnaturally quiet Portuguese "mouse" was reverting to a more naturally ebullient, sunny person who was finding science exhilarating. Her English was now fluent; she had learned to queue; she had learned how to run for the bus; she was, in fact, at the beginning of a deep love affair with life in England. With a friend she bought an old fox-hunting horn in the Portobello Road and they blew it in Hyde Park. She lived in a "bed-sitter" in Golder's Green and would take the small Green Line bus to Windsor to have tea in a little tea shop. She learned much scientifically but much more humanly. The experience was "like going home." There had been so many things she had found disturbing in Portugal, but—never having lived anywhere else—she could not tell if her intuitive judgment was fair. Then she came to a country where people lived so differently and behaved so differently, and did all "those marvelous things such as queuing quietly, and respecting each other, and not minding one's house being untidy, and going into a friend's house spontaneously without the friend's feeling obliged to tidy things up." In sum, she found surcease from all the small obligations that she had always regarded as standing in the way of simple friendships and true human relationships. Above all, there was the English countryside, with its green fields and its peaceful light, the incomparable light of southeast England.

She was truly happy—and one thing only remained. As Anna still insists, a virgin discovery becomes a real discovery only on the day that two people *genuinely* believe in it. It was to become thus real only in the spring of 1965, on the morning when Dr. Vera came to the laboratory upstairs, sat down, and said, "I have been invited to give a talk at the next meeting of the British Society of Immunologists, and I would like to present your stuff. May I have your permission to do so?" She continued, "I am going to call these areas 'thymus-dependent areas,' " the very phrase Anna had been arguing about for six months.

Dr. Vera's paper was duly given and Anna was present, and the results of Anna's hours at the bench and first experiments finally saw the communal light of day one year later, in the *Journal of Experimental Medicine*, on October 22, 1966. By then she was a scientist in Dr. Vera's definition of the word; she had learned the fundamentals.

When I asked, "What are they?" she replied, "Well, there's creative observation for a start. You see, I was reading a lot. The National Institute has a wonderful library, so I began to read a lot of the old stuff, seeing what people thought before, seeing how wrong they were by being straightforward morphologists. Just looking at things is terrible. You start thinking what you see, and believing what you see, when you should be doing experiments to *prove* what you see. So there's the designing of good experiments, too. Then there's the capacity to think up concepts to explain the patterns you are seeing. Above all, there is freedom, the knowledge that you are going to make mistakes and not being afraid. To be frightened of making mistakes is to be in prison. By the end of that time I was dropping mistakes right, left, and center. I learned to love making mistakes.

"But the most fundamental lesson of all is to learn to know what you don't know. Because then you are free to play with the little you do know. You also learn what you cannot do—what your limitations are: thinking limitations, hand limitations, feet limitations, whatever limitations you like. So then you know your assets, and finally when you have a fairly critical view of these, you don't waste time and energy. Up to that time in my life I had wasted an awful lot of both."

According to Dr. Vera, Anna had demonstrated additional qualities by then: tenacity, imagination, stubbornness, confidence in herself, and reaction against assault, the reaction that says, This is my territory, and no one is going to tell me that I shouldn't paint my front door blue. These are, in fact, some of the qualities which ensure that a person is more than just a humdrum scientist. Being good at looking down a microscope or scanning the literature or carrying out an experimental routine—thus showing that one can see, and understand, and manipulate—add up to competence. But once one has made discoveries the bells start to ring; there is a feeling of resonating with the world; one knows that one has arrived: it's happening, it is here. The phenomenon one is grasping at and understanding is something quite new, something very special. It is something that is not going to be allowed to get away.

Yet one more step must be taken—one vital to the whole scientific ethos and crucial for the individual and the enterprise. Without it no one could ever tell for sure whether our ideas of nature are true or not. It is, of course, the very necessary act of exposing one's ideas to personal and communal criticism. Such criticism can, when used indiscriminately or absolutely, be destructive of both people and ideas. This essential act of constant exposure prevents many people from becoming scientists or remaining scientists, for one must have vast reserves of lonely self-confidence to sustain oneself in the face of pervasive collective criticism. Anna perhaps showed one sign of newly achieved, and accredited, scientific confidence. In the spring of 1965, after her paper to the British Society for Immunology, Dr. Vera asked if Anna had approved her presentation of the data. Dr. Vera had given the paper very competently, although, she said later, somewhat nervously, because she knew she had left out one of the points her colleagues had badly wanted made. Still, she was a little taken aback when Anna replied that while the presentation wasn't all that bad, it wasn't as good as Dr. Gowans's!

CHAPTER 4

TWO LONG
NIGHTS

Thought is only a flash
between two long nights,
but this flash is everything.
JULES HENRI POINCARÉ

March 1966 marked the end of Anna's fellowship and assignment abroad, and in a welter of unhappiness she took the boat back to Lisbon. "Oh, that bloody boat," she recalls. "I cried so much." Her parents came on board with the pilot, and she remembers wanting to see them and hating to see them, knowing that she didn't want to be there at all. But at least there was the brave new world of science to build.

The moment she disembarked she went straight to the laboratory at Oeiras. It had been built in the gardens of the palace of Pombal, owned by the Gulbenkian Foundation. A small river runs through the gardens, and the surroundings are extremely beautiful. Lisbon, built on more hills than Rome, with a superb river that widens into an estuary and the far reach of the Atlantic, is a wonderful city. The lab, too, was magnificent. Every day in the sunshine she would drive there from her home, along the shores of the Tagus. In the winter the staff would eat lunch in the canteen in the cellar of the old palace, but in the summer they picnicked on the beach. Everything was beautiful, she recollects, but she was unhappy.

Her first interview with her new boss was disastrous, a putting down of all her work. The tentacles of the insular and self-important regime had not loosened one iota, and the scientists were afraid: afraid to make mistakes, terrified to take the initiative, scared, above all, of people whose new ideas might challenge the status quo. So they neither believed her achievements of the past two years nor wanted to hear about them.

In the summer, Dr. Vera came out to Portugal. At that time, Dr. Peter Medawar was doing work on antilymphocyte antiserum (ALS), a fluid that kills lymphocytes, and he sent along a small sample of his material. Dr. Vera's official reception in Lisbon was about as cool as Anna's had initially been in London, but together she and Anna did some of the first experiments showing the effect of this antiserum on the lymphoid tissue.

It was a time of great tension. Anna was contemplating marriage and was generally expected to assume a woman's traditional role. The conflict between these claims and those of a concentrated scientific career strained her to breaking point; the crisis came when she and Dr. Vera had an icy row right in the laboratory because Anna was planning to take a very aged aunt for an afternoon ride in her car. Suddenly Dr. Vera shredded her to pieces, urging on her the obligation of getting away from mediocrity and the debilitating confinement of her current existence. How *dare* she waste her talent?

By then, of course, Anna respected Dr. Vera's judgment highly. If Dr. Vera told her she was very intelligent and could be a good scientist, she was prepared to believe it. But she also felt the weight of old pressures. (And it *is* important, she still says, that, in a small group, younger people look after the old, are kind to them, and make them feel they have a reason for existing.) The issue was, of course, where her ultimate allegiance was to lie. Fiercely, Anna defended her own stance and her loyalties toward her family, but from that moment it became apparent that sooner or later she would leave.

It was the inevitable end of a cumulative process, for it had also become clear that scientifically they were not going to get anywhere in Portugal at that time. She had tried to persuade her colleagues that new discoveries had launched a period of great promise and excitement in immunology, but she finally understood there was no point in saying even that. She began to look forward to the day when she would go into the office and say, "I'm leaving." Still, it could not be a matter of just opening a door, walking through, and shutting it behind her. In November 1966, she became seriously ill. Medically her illness was not well managed, and she nearly died. But she recognized quite clearly the nature of the psychosomatic process that was going on, and the experience was both consoling and maturing at the same time. She also saw that her illness could justify her departure, because, as friends were quick to point out, she would be so much healthier elsewhere.

She decided to pause in London on her way to give two papers at a conference in Holland, and before leaving she presented the papers as a trial run in Lisbon. The effect was crushingly decisive: "I don't know what you are going to talk about in Holland; I don't see the relevance of anything you say; I don't understand what all the fuss is about. There is nothing."

In London, Anna consulted a neurologist, who discovered that the many steroids and drugs administered by her doctors had taken a severe toll. But even so, she recovered rapidly, and when she learned that a lectureship had become available at Glasgow University, went up to Scotland for an interview. To her delight and amusement her interviewers seemed much less concerned with her qualifications than with the possibility that she might miss the southern sunshine in that bleak northern city. She was back in Portugal when the Glasgow appointment came through, and she had the pleasure of going into the lab office and saying, "I'm leaving," and then the pain of saying the same thing to her parents.

The years 1967–71 Anna remembers as a blur, though scientifically they were, she says, years of transition from sloppiness to precise definition. Several things evolved from the work she had done with Dr. Vera: First, the structure of the lymphoid tissues in the mammal was finally worked out; second, it became clear how the two populations of lymphocytes—the one thymus-dependent and the other independent—fitted into this structure; third, it was reconfirmed that these two populations moved to their own specific sites in the spleen and the lymph nodes. (Scientists call this "homing.") Fourth, Dr. Vera's hypothesis—that in order to mount an immune response to certain foreign substances, both populations of lymphocytes are needed—was also confirmed. By 1971 all the real groundwork had been done. Many scientists had seen the facts and had conceived new experiments that extended from the initial observation. So a small invisible college began to form, of scientists bound together by a common curiosity about the subject of lymphocytes, their movements and populations.

In Glasgow, Anna began her Ph.D. thesis on cell traffic and the patterns of cell migration in animals, and this work forced her to sit down and, on her own, think out all her ideas and their implications. The thesis developed and extended her original discovery, that the lymphocytes are not distributed randomly in the body but are specifically organized and when they enter the bloodstream, go mostly to the spleen and lymph nodes. There, once again, they organize themselves into specific zones—as groups of cells that have somehow recognized each other and belong in the same area. This was Anna's finding, and she gave the phenomenon a name: *ecotaxis*, from the Greek words *oikos* meaning "house" and *tassein*, "to arrange." Then, having assumed this "self-arrangement" to be a basic property of lymphocytes, she was, of course, obliged to try to explain it.

First it was necessary to find out if this property was widespread in the animal world. Does it occur in *all* animals with an immune system? Does it occur in any other mammals besides mice and men? Is it a genuine phenomenon in other species? When did it first begin in evolution, and why did it occur? Collaborative experiments on chicks and on fish had demonstrated that in these species lymphocytes do behave as in mammals. But were there comparable properties in the cells of organisms much simpler than fish? Could any light be shed on ecotaxis by studying these simpler cellular systems? If the tendency of lymphocytes to move as if they knew where they were going, and to stop in well-defined zones, *is* a fundamental biological property, then a study of simpler systems should reveal the precursors of this pattern, and a detailed study of these simpler systems might well throw light on the question of "how," physiologically, this precise positioning occurs. She had to know if a mature lymphoid cell *knows* where to go and so goes straight there regardless of any other cells it meets on the way, or if its final position is simply the end result of a series of interactions with its neighbors, a consequence of many casual encounters rather than a predetermined event.

At this juncture Anna badly needed the wisdom of a cell biologist. She was fortunate when, early in 1971, the department of cell biology at Glasgow University held a seminar on the "sorting out" of cells. Anna attended, and when later she organized a conference on the traffic of lymphocytes and other cells, she asked Sebastian, the professor of cell biology, to join in. Jim Gowans, who had been the first person to discover the circulation of lymphocytes, came along too, and everyone attacked the basic questions: Why do these phenomena happen at all? How do the cells know where to go? How do they know when they are "there"? Why do lymphocytes sort out into precise places in the organs? Why and how does lymphocyte traffic occur at all? It was a very lively affair.

As a prospector may sense that "there's gold in them thar hills," so one scientist may sense that another colleague has a mine of knowledge that will be valuable for his or her problem. Though Anna knew little in detail about Sebastian's work, she felt intuitively that it could shed light on the problem of lymphocyte movement and positioning.

While at Cambridge, Sebastian had intended to become a geologist, but after he took up diving, marine life proved to be far more interesting than marine sediments. So he began to read all the classical zoological papers on sponges, those very simple animals whose varieties range from single-celled organisms to the vast colonies whose skeletons make up our coral reefs. Sponges beckoned enticingly; so too did the possibility of trips to Bermuda or the Bahamas to study them. (Regrettably, these exotic journeys never materialized because British sponges, and especially Scottish sponges, always proved good enough for his purposes.)

What makes the cells of a sponge stick together? It's an old question. In a classical experiment, done two hundred years ago, a scientist named Abraham Trembley had forced a whole sponge through a very fine gauze into a container of seawater. The cells of the sponge separated completely from one another, but after twenty-four hours the sponge had reconstituted itself. Clearly, even at this very simple level, there are recognition systems that tell a cell, "This one is like you, so join up." Perhaps, Sebastian had speculated, cells produce factors that promote adhesion and others that inhibit it.

It turns out that there are indeed substances that keep sponge cells apart, and similar factors exist, Sebastian has found, in chickens, in horses, and in cows. These substances are necessary because cells are quite indiscriminately sticky objects and therefore, if you want one specific cell to stay alongside another specific one, you have to produce something like an "anti-glue," which will keep the cell from sticking to everything else around. One wants liver cells to stick only to liver cells, and the reason we don't get lumps of liver somewhere in our brain is that factors are produced which keep differing cells apart. Should a liver cell bump into a brain cell it would move away. Cells are naturally mobile, and as long as they receive the message, "Go away," they will keep on moving.

Sebastian extended his work to lymphocytes because "Anna pestered me." She pestered him because she felt that biological principles similar to those at work in sponge cells might also be at work in lymphoid cells. Quantities of lymphocytes would regularly arrive in his department with a message that said, "Please measure their adhesion." As more and more lymphocytes came in, the technicians used to run away in dread, and Sebastian confesses that there were times when he nearly threw them all out.

He himself was only marginally interested in the lymphoid system. He knew absolutely nothing about lymphocyte biology, and at first he didn't feel that it had any relevance to basic cell biology. But the moment he realized that the very same factors that kept sponge cells apart might also keep the T- and B-lymphocytes in separate places in the spleen and lymph nodes, the question became thoroughly tantalizing, experimentally and intellectually. Finally he did measure the adhesion between the lymphocytes—as asked—and then he and Anna undertook a whole series of experiments which demonstrated that, just as with the cells of sponges and chick embryos, mammalian lymphocytes produce substances which *do* diminish their "stickiness," both to different types of lymphoid cells and also to other, altogether different, cells.

In 1971 the two scientists showed that T- and B-lymphocytes *do* have different adhesive qualities and thus do influence each other's positions. That finding led to further experiments in the three years between 1972 and 1975, undertaken to discover whether or not the actual circulation of lymphocytes throughout the body is mediated or modified by interactions between lymphocytes and other cells in the tissues. The work was carried out in collaboration with a small, brilliant bundle of bearded self-assertion, Angelo, a student from Lisbon, who, on hearing Anna lecture on immunology, became interested in the problem, and decided to do his Ph.D. with her in Glasgow. As is usual with creative people in science, he rapidly found his own variations on the basic problem. Being, Anna says, "a very creative chap," he tried, in a series of beautiful experiments, to modify the membrane of the lymphocytes to see if their movement was affected. It was.

These years were really crucial for other reasons, because it was during this time that Anna first suspected that the development of certain diseases might be due to, or closely connected with, trapped lymphocytes. Now it was that she first made explicit to herself the question that was preoccupying her when we first met and talked about Hodgkin's disease.

In the spring of 1972, she was asked to give a talk to the Immunological Society of Portugal on immunodeficiency. Just before she left she received her weekly subscription copy of *The Lancet*, a British medical journal, and in it she read an article about patients with severely impaired immune responses and yet another form of enzyme defect. And she decided, To hell with it. I'm not going to review everything that is known about immunodeficiency. I had better sit down and think. As she did so, she speculated: Might there perhaps be some forms of human disease marked by an immune deficiency reflecting not a quantitative presence or absence of lymphocytes in the body but, instead, the abnormal *movements* of these lymphocytes? In quantitative terms, the sum total of the lymphocytes in the body would be all right, but they wouldn't be in the places where they ought to be, or might need to be. Suppose, for some reason or other, they were hiding, or trapped, in some organ or other; in that case, they would be neither in the circulation nor in the tissues where they were needed to deal with infections or foreign agents. Indeed, they might not be able to get there. If this were so, the usual diagnostic index of immune deficiency, the total quantity of white corpuscles in the blood, would be a red herring. It would not really reveal what was going on; the clinician would be misled—looking in the wrong place—and the trouble would be elsewhere. In her paper in Lisbon, Anna proposed this as an idea, and she also suggested that one of the hiding places for the lymphocytes would be the liver. Later in 1972, she gave the same paper to her colleagues in Glasgow.

At that stage, what eventually was to be a discovery was just pure thought. The image she consistently refers to is "working within the deserts of thought, like those fantastic spaces in the pictures of Max Ernst or Salvador Dali." There is emptiness, then suddenly there are signals. There were to be two such.

The first occurred during her visit to Portugal. A pediatrician who came to the lecture, the husband of one of her students, said during the discussion, "I have an odd observation. I have this child with Hodgkin's disease, who was given a certain treatment. Then he caught measles and suddenly he was well again." Now at that time, Anna believed that Hodgkin's disease was caused by an *absolute* deficiency of T-lymphocytes, because quantitative blood counts showed that there really were fewer than normal in the patients' blood. She also knew from past studies that measles too affects the T-lymphocytes and diminishes the immune response. If one were thinking along traditional lines, measles added to Hodgkin's disease should make things immunologically much worse. Yet the pediatrician said they were better, and thus his observation rebounded smack against the wall of accepted belief. This small piece of empirical information, which in fact other people had both noticed and remarked on, was recognizable as a crucial signal simply because it could now be explained according to what Anna was currently thinking. When she gave the same paper in Glasgow, she raised the issue once again: If the patients sometimes get better— throw off *both* their measles and their Hodgkin's disease—as clinicians know they sometimes do, then, of course, all the lymphocytes can't have vanished; they have to be around *somewhere*.

This thought had never really surfaced until the day she was talking to a colleague, Dr. Hami Singh. They were discussing leprosy, a disease in which, as in Hodgkin's disease, lymphocytes seem to be missing. When leprosy begins, and as it progresses, the lymph nodes empty, and this fact has again been interpreted as indicating a lack of lymphocytes. But when leprosy is properly treated, all the lymphocytes reappear, returning to the lymph nodes just as though the migrating season were over. This was the second signal, a most striking and clear one, because it *showed* that the lymphocytes must have been hiding or have been trapped somewhere.

The hypothesis which was triggered by these signals was like the idea of zero for Anna. "Unlike my first discovery, which was an observation, this idea was pure thought," she said. But, of course, in science thoughts must be turned into data, into detailed results, to be tabulated and plotted. Eventually one hopes to have a graph whose patterns will indicate whether or not the idea might be right. It is going from mathematics of one sort—an abstract concept—to mathematics of another sort—a graphic representation that demonstrates underlying patterns instantly and in an easily understood form. To achieve that further step, Anna needed time and patience. First of all, she had to demonstrate unequivocally that in these diseases, Hodgkin's disease and leprosy, the lymphocytes do *not* vanish but are simply trapped. To show that, she needed access to good clinical material, and the best place for this was America. In 1974–75, when her sabbatical leave came up, she left for six months' work in New York. That was when we met.

CHAPTER 5

RANGING

Creative imagination must stop
well short of delirium.
CALVIN WELLS

With easy access now to extensive clinical material, thanks to the close connection of a cancer hospital with research laboratories, Anna began a series of clinical and experimental studies in an effort to establish whether or not there was preliminary evidence of the maldistribution of lymphocytes in Hodgkin's disease patients. Was it even reasonable to suppose that diseases could come about because lymphocytes were delayed or trapped and therefore not available to do a proper job of surveillance? It was a novel idea, certainly, and as such could be expected to run into problems on that ground alone. But was it a likely idea or just a ridiculously implausible one?

She voiced this one evening shortly before going to give a lecture at Harvard Medical School. Places like Harvard were scientific sanctuaries, she insisted, but she did not mean a place to flee to when you are persecuted. They were places to go in order *to be* persecuted. A piece of work that is entirely new, which may perhaps open new doors onto the world, must be seen by the people who can destroy it best. If you are not prepared for a rough time, to be attacked by your scientific peers, then you might as well sit tight in a little corner, saying nothing.

Whenever a scientist comes up with a new idea or technique, most people don't understand it at all. Only those with very fresh minds can hope to. At our first meeting I was able to grasp her ideas with sympathy only because I had no preconceived notions about Hodgkin's disease or lymphocytes and no emotional commitment to any particular theory. By contrast, incomprehension is an occupational hazard for mature scientists, whose mental compartments may be so full of their own understanding and their own ideas that they can't appreciate anything else—indeed, may not want to perceive anything else.

Anna kept re-emphasizing the notion of "space." There is a limit to the number of people who can occupy a specific physical space, and it is just the same with ideas. Accepted ideas tend to fill all the mental space available. In order to make room for new ideas, you may first have to clear out the old, and a scientist with a message treads a very fine line indeed between saying the same thing too often and not saying it often enough. Hence, according to Anna, it can be terrible to make a discovery when you are young.

"Why?"

"First of all you may become complacent. Then you may be doomed—no, not doomed, but haunted by your own ghost, repeating the story over and over. So then you run the risks of timing, taking enough time to make people understand and in the end talking about it for far too long.

"But one day the point comes when you *know* that your discovery is in the collective thinking. You may be giving a paper, and suddenly you will see people in the audience becoming more and more bored. And they are bored because by then everybody knows what you are saying, accepts what you are talking about, and wonders why the hell you are taking so long to say something so obvious.

"After I had made that observation of the positioning of lymphocytes, I had to sit down and think, for my thesis, about the way the cells arranged themselves in compartments. And I gave the phenomenon a name: Ecotaxis. From now on I shall be talking to everybody about ecotaxis, and one day people will be bored, because they'll feel it's been known for a long time. But the problem of ecotaxis is: Why does it happen? How does the lymphocyte know where it is and when to stop moving? What captures it? That is what I'll have to answer next."

I objected. "How do you know that it 'knows' where it is? You immunologists keep talking about 'homing.' The lymphocyte seems to me more like a migrating cuckoo than a homing pigeon. The difference between a homing pigeon and a migrating cuckoo is that the first bird doesn't know where it is but knows where it wants to go, while the second bird knows where it is, but doesn't know where it must get to. You take a homing pigeon in a railway carriage, and somewhere along the way you let it go. It doesn't know where it is but it does know in which direction to head, and when it gets home it will recognize that *it is* home. A migrating bird *does* know where it is, but if it is a young cuckoo that migrates after its parents, it doesn't know where it is supposed to be going. Yet somehow it must recognize when it has arrived because it stops. Now a lymphocyte seems to me just like a young cuckoo in September. It can't possibly know that it has to go to the spleen or the liver. And therefore, all the time you scientists have been talking about the lymphocyte 'homing' to predestined sites, I think you probably ought to have been using the word 'migrating.'"

Anna took an enormous breath. "I've been for ten whole years trying to get someone to put it just like that. That is why I invented the word 'ecotaxis,' because homing is strictly for pigeons! And it isn't like homing. It is like migrating birds. That's the fundamental difference, and that is why I thought it necessary to have a new word.

"You see, the essence of the phenomenon *is* the relationship of the cell with the environment that stops it, that holds the cell there. O.K.? And that is no different from the relationship of the migrating bird with *its* environment, that holds it there. But much too little is known about this at the cellular level. You've made me think of one thing that would be worthwhile doing—something as simple as measuring the temperature of different areas of the spleen, and so on. Suppose there are 'southern Spanish regions' in the spleen?"

"But lymphocytes don't go to the spleen to reproduce," I said.

"But oh, you are wrong," said Anna. "They *do* reproduce in the spleen, and they reproduce in the lymph nodes too. Of course they do. When they get a little stimulation, they divide and divide. You have to have an antigen or something coming along to start them off, but that is where they reproduce."

"So perhaps lymphocytes do need a warm climate."

"Wouldn't that be fascinating?" said Anna. "But we'd need a new technology. We'd need new tools, like thermometers for measuring right inside the centers of organs—into the cells. We need new tools anyway. It is going to be so exciting, the next twenty years."

It was exciting the next twenty days. Calling me in one evening toward the end of her sabbatical leave, Anna announced that she had finally got some results together. It was, she said, a "keyhole" time. She had a small shaft of illumination, promising and tantalizing, but there was a lot more to Hodgkin's disease than anyone had thought. At an early stage in the disease, she reminded me, only the lymph nodes are affected. But as the disease progresses, the spleen may be involved and must be taken out to see whether or not it too is diseased. Unlike other organs, the spleen is rarely biopsied because it is permeated with blood vessels and so full of blood that the patient might bleed to death if it were cut into. Hence, in order to determine the extent of the illness, and therefore the therapy, the spleen has to come out. This diagnostic procedure proved to be a stroke of luck for a scientist who wanted to see whether or not lymphocytes were trapped in the spleen in Hodgkin's disease. Counting the lymphocytes in the blood taken from a vein before and after the operation, and also in the spleen itself as soon as it was removed, was a relatively straightforward matter.

In this study, she was collaborating with Dr. Tien-Chun, a pediatrician, who at the time of Anna's sabbatical had a small group of patients in Stage I of Hodgkin's disease. Every patient had been thoroughly "worked up," meaning that as much data about them as possible had been obtained from a whole range of biochemical tests; thus blood and lymphocyte counts were readily available. Some patients had not yet had their spleens removed, and the graphs showed the curve of the lymphocytes in their blood dropping down and down with time. When given the same range of tests after the spleen was removed, the lymphocyte count either remained the same or unashamedly went up.

In the case of one patient, a little boy who was at a very early stage in the disease, the first studies showed that the number of T-lymphocytes in his blood was very low: 23.5 percent at the time his spleen was removed. Normally there are 70 to 80 percent. When the spleen had been removed and the percentage of T-lymphocytes within it immediately estimated, the count gave 61 percent. Normally there are only 30 to 34 percent. Once the spleen had gone, the number of T-lymphocytes in the blood increased by 50 percent, and it stayed around that figure. Had some T-lymphocytes been sequestered in the spleen? And were those trapped lymphocytes—if indeed they really had been trapped—living, functioning cells? Studies were done before and after the spleen removal specifically to test the T-lymphocytes in the blood for vitality; then those T-lymphocytes that had been separated from the spleen were similarly tested. All were exposed to the substances I had met the first night in the laboratory, the mitogens, that stimulate cells to divide, and their dividing activity was carefully measured. Though the vital response of the lymphocytes in the blood was low, for those in the spleen it seemed normal. It was thus possible that fully functioning lymphocytes *had* been held in the spleen, and that when it was removed some degree of normal immunological function was restored because the lymphocytes were no longer trapped there.

From the point of view of Anna's hypothesis this particular patient was a neat textbook case, and the results were a first indication that concept and data were possibly beginning to converge. But it was also made quite clear to me that what had been one idea, or one question, was rapidly evolving into many questions that had to be asked and could be answered.

One such, emerging out of the figures and graphs that incorporated the data on other patients, was the following: Was the process the data reflected what normally happens after a spleen is removed, whether it is diseased or not? Would one find the same effect if, for example, one could do the range of tests, before and after, on a healthy person whose spleen had ruptured in a traffic accident—a not uncommon occurrence? Accidents are one of the most reliable sources of normal spleen material.

There were other questions that had to be raised: Did the spleen act as a lymphocyte trap in normal conditions? And were there, perhaps, two distinct populations of T-lymphocytes, one generally to be found in the blood and another susceptible to trapping in the spleen? The graphs also showed that there were changes involving populations of B-lymphocytes too, and these were a puzzle. Anna could not fathom them. Certain of the results seemed to indicate two other possibilities that had to be considered: Either the T- and B-lymphocytes were interacting somewhere other than in the spleen or some other as yet unknown process was going on in Hodgkin's disease which was affecting the B-lymphocytes as well as the T-lymphocytes.

That evening was my first introduction to the range and depth of science. The initial question—In Hodgkin's disease are lympho-cytes trapped in the spleen?—might sound simple enough, but from then on we would be in a maze all the way. This was not just a sim-ple matter of counting lymphocytes. The leitmotiv was complexity: complexity of painstaking experimental and surgical techniques; of repetitive tests, each of which would reveal one small facet of this living microcosm; of time-consuming procedures, not a step of which must go wrong; of machines for counting, spinning, blending, and seeing which must not fail; of reagents and substances that must be pure; of temperatures that must not change; of figures that must be accurately assessed; of points on a graph precisely plotted, and curves correctly outlined; of deductions that must be cautiously drawn. Every step of the process must be impeccable—not so much foolproof as scientist-proof. For no one should be able to punch holes in any part of the fabric. On the contrary; once the results were eventually committed to paper, step by step, other scientists should be able to follow the whole procedure until the final deduc-tion seemed a valid one to them too.

Thus Anna cautioned me that she really should not be talking about these results yet. It was not fair to talk, even in principle, about results that one had looked at just two hours ago.

"Not fair to whom?" I asked.

"Not fair to the results themselves," she replied. "Because one has not seen enough. I need many more patients, many more fig-ures. We do have some more results, and I shall see them tomorrow. But the only reason I am talking at all is for the historian. Then you can record the misery."

"Is the unexpected the misery?"

"No, not at all," she replied. "The unexpected is never a misery. It is the expected, the monotonous and the immediately understood, that is a misery. I am perfectly ready for the unexpected, but . . . no, what you are really getting me to talk about is my self-doubt. *This* is the misery. Is this all real? Now that I'm going to give big lec-tures, do I, in fact, have anything to talk about? I think I do, but this is a terrible moment, of total self-doubt about the relevance of all this lymphocyte traffic business."

"So far as Hodgkin's disease is concerned?" I asked.

"Not only so far as Hodgkin's disease is concerned, but so far as everything is concerned. I'm absolutely full of self-doubt."

"But is this the first time you have had such doubts?"

"Yes, and those empirical tests only stress it."

"Why?"

"It is because they are so beautiful. The results are so beautiful, confirming what I thought, that I can hardly believe them. This is when one must doubt very, very fundamentally. Let me show the graph curves to you again; then you can see for yourself. This is a Hodgkin's disease patient, in Stage I, before splenectomy. That top line is the whole blood count. Then this line is the one with the wrong things, and here is one that shows the lymphocytes and the monocytes—the macrophages of the blood—and as days go by these are dropping, dropping, dropping. Now look at this other patient who has been splenectomized. Immediately, the whole white-blood-cell count goes up. The one before splenectomy is going dramatically down and this other is going dramatically up. So there seems to be no question that the spleen is trapping the lymphocytes."

"So it fits," I said.

"Absolutely. Now look at this next patient—one that was splenectomized a long time ago. Here you can see that after the operation things seemed all right for a while. But then, as the days and months go on, the counts begin to go down again; the lymphocytes are being caught again in some other place.

"Let me show you even more curves, also taken over a long period of time, because then if you ask me a question as to what's really going wrong, I can perhaps answer it in terms of the cells. Look: In these graphs the Ts and Bs are shown separately. This is a patient who had *not* been splenectomized, and here is the line of his lymphocytes—both the Ts and the Bs—going down. Look at the curves carefully. Now there at this time is the operation. Right after the splenectomy there is either no change or the Ts go up. But after a while, the Ts again begin to go down, but the Bs still go up. This is why we need more points in the graph, more patients. This is why I keep saying, Look at these two patients. If you keep the spleen in, the T-lymphocytes go down. If you take the spleen out, they go up. But because I believe that Hodgkin's disease happens because—or tells us that—something is fundamentally *wrong* with the T-lymphocytes, *eventually perhaps they are bound to go down anyway.* So, what do these results really show me? That the spleen really does trap the lymphocytes? Or only that there *is* a defect in the T-cell, and no matter what you do when treating the disease, the number of Ts is going to drop all the time?

"But my self-doubt really comes from the fact that actually it all is so much what I would expect. So far as the Ts are concerned, they are behaving exactly as I would predict. They are behaving so neatly that I'm beginning to have terrible doubts about the whole hypothesis."

"But why?" I asked. "Many people would say, 'How clever I am! What I predicted came true!' If it seems that good, why don't you just accept it?"

"There are many reasons. One reason, of course, is the scarcity of the numbers. At the moment we have results from only five or so patients—which is quite good for Hodgkin's disease. But I need more."

"You'd be happier with twenty-five?"

"Naturally. I'd be happier with fifty-five. But the other reason is that if I discuss the results with the people helping me in the lab they will believe anything I say, and that will multiply my self-doubt threefold. I must discuss these results with an equal. I must discuss them with Marion McLeod [a Glasgow colleague in New York that week]. I must discuss them with the director here. I need to argue with someone who will look at these results and say, 'Rubbish, you've got nothing here at all. They don't fit any theory whatever, least of all your own. This is absolute rubbish!' I need somebody to tell me that very much."

It was essential, Anna emphasized, that she remember constantly her limitations as a scientist, and especially when a prediction comes true. She uses her eyes well, but she is not so expert with her hands. She is unusually single-minded, and that is both an asset and a limitation. It is an asset because results which have been predicted and then found can very easily be fitted into a theory. But single-mindedness can be a limitation too, because for every set of data there can be at least ten different interpretations which might be obscured by a single devotion to one.

"The curves are here," she continued. "Actually it is incredible that they exist at all. Blood cancer cases get so complicated, especially after drug treatment, that you rarely get patients in a pure situation. But so far, nothing therapeutically—but nothing—has happened to these people. Short of taking a spleen out of a normal human being, before and after doing peripheral blood counts, which is something we will never be able to do, it is the clearest situation for me at the moment. So I must not believe it. Or if I do believe the data, I must try to see them from the point of view of many other interpretations."

Over the next few days the data took possession. She worked for hours with the graphs, going backward and forward with the figures and curves, not only those incorporating the clinical data of the patients but also those from her own laboratory studies of their blood and lymphocytes. Science is the polar opposite of dilettantism. One can speak of a person being possessed by the data, but this concentrated restlessness was more like being possessed by demons.

There was a second reason for her preoccupation, which I learned of one day when we were about to cross the street. Anna, watching the traffic lights, held me back and then charged off across the street, only to tease me later about my reflexes. Crossing the street, she said, was like measuring up in science, the difference between people who have quick reflexes and know when to move and people who do not. Quick reflexes can be dangerous in real life, and equally so in science, but having them can also pay off. Vision in scientific life really means much more than just seeing things faster than other people. There is nothing special about *that*. But both the speed with which you appreciate the essence of a new scientific situation and the speed with which you exploit it are what distinguish one scientist from another. Whether empirical facts or new patterns, the signals coming from the desert may be impinging on two scientists simultaneously, but one will respond far faster. Scientific reflexes genuinely exist, and hers were telling her that she might have to leave Glasgow for good.

To a degree she had known for a long time that this must happen. At international conferences in Dubrovnik and Glasgow, Anna had realized that in America were the clinical material she would need and the standards she had to match. Furthermore, the Portuguese revolution of 1974 had reawakened her dormant social concern, first triggered by the fisherman of Nazaré, and she no longer felt able to pursue lines of pure thought only. She would have to try to make a more direct impact on health problems. She had considered returning as a doctor to the poor rural areas of Portugal and had even mentioned it to the ex-minister of education. But he, with a deep understanding of both the passive village and the dynamic woman, had urged her to stay in research, saying, "Spare the village, I beg you! Spare the village!"

Now, again, she would have to break away and start all over, and she both embraced and resisted the impulse. She felt that she had reached a plateau of happiness in Glasgow that she was unlikely to regain for a while, and she wanted more than anything to remain there. But she was not going to. By the end of her sabbatical leave she had decided she would move to America, and would do so the following year.

Scientifically it was probably a wise decision in more ways than one. Over a period of time some scientists seem to lose the ability to make discoveries, not perhaps because they really have lost it but because they have allowed themselves to be swamped by a host of other things. For Anna to nail down her hypothesis—that certain forms of immune deficiency can result from lymphocytes' being trapped—and to find out what causes the trapping would take five or seven or even ten years. And in 1975, her main asset was that she came to the clinical situation and its data with a completely fresh mind. Ten years later on, the theory either disproved or totally confirmed, her mind might well be cluttered. At the start, all she could honestly say was that she had the first evidence of a possibility that raised her old questions in a new, tantalizing form: Why do the lymphocytes arrange themselves in the organs, and why do they sometimes get trapped there?

She returned to Europe in early April. Before she left we discussed the experiment she had been doing the night I sat late with her in the laboratory. It had gone along fine, but finally the time came when she had to take the materials out of the cultures. And, she confessed, "I nearly messed it all up. Because all the time I was writing a poem in my head, trying to express what we had been discussing about collective and individual action and what will endure."

In the final week her doubts about the significance of her results rose up once more.

"What is the real measure of an event?" she asked me. "Of the worth of any individual event?"

So far as the scientist is concerned, the answer, we decided, probably is not its mere occurrence but the number of things that follow because it has occurred. The essence of the individual contribution and its significance is that one should produce something—concepts and results—that can merge with the collective experience. And the measure of the individual is in the brilliance, the ease, and the humaneness with which that is achieved.

CHAPTER 6

IMAGININGS

What is now proved was once only imagined.
WILLIAM BLAKE, "The Marriage of Heaven and Hell"

The months between April 1975, when Anna left New York at the
end of her sabbatical, and the end of April 1976, when she returned,
were not conducive to tranquil research. In addition to the thousand
jobs to get through as one home and one laboratory were wound
down and others started up some 3,000 miles away, there were experiments to do, students to teach, examinations to mark, and theses
to supervise. We corresponded frequently (we also spent three great
stretches of time together), and the flavor of that year-long interval
and her first three months in New York is captured in Anna's letters.

Glasgow
April 18, 1975

*I listened to one of our tapes primarily to check my accent, and I
was appalled that for the most part I "sink" rather than "think."
What does it mean to be Portuguese, you once asked me? Carrying
the burden of the "ç"—cedilhado—(almoço, lança, laço, praça, etc.,
etc.) in my çoughts.*

April 22, 1975

*The abstract is written. On to the next task of starting to work on
the New York results and papers. In this quiet house all that grows is
invisible, except the odd readable assertion. I love it: its silence, its
filtered colors through gray skies. My upstairs neighbors—crying*

gulls—are on the roof in the sky, singing their heads off, making a world of sound that is very reassuring in a gray city. It is good to be back.

April 23, 1975

I sat going sadly through a book of Whistler paintings, looking slowly, page after page after page. As if doing things slowly could make time itself go slow, and the moment of departure delayed—delayed to the point of never having to happen.

How the living of this time to come is going to be lived slowly—like eating slowly, or walking slowly, or even listening slowly—so as to create the illusion of distance from the end, from an end that is near.

From an end that, in truth, has been here for a long time. At an international meeting on immunodeficiency in Oporto, years ago, the quality of work being done in America hit me, like a splash of cold water on a hot face in the hot sun. I remember sitting and listening to the speaker and wondering what we were doing in the modesty of our limited clinical setup.

Then when some American friends popped into Glasgow on their way to Edinburgh, I remember crying silently in an empty cab after leaving them at their hotel. I knew then that all these gentle English ties were sooner or later going to be severed. And I would go on alone, without the extreme simplicity of my quiet English life.

In my mind I hold a strip of images that contain my aged parents at Lisbon Airport, saying a sad goodbye, and the glory of the English green by the English sea, and I have almost the feeling of being cheated again. It is painful to depart. It is always painful to depart—even if I do not go. I have found myself moved to tears going to the quay in Lisbon and watching people go, and others crying at the separation.

I've always said that I miss things more than people, which is true, and for which my mother thinks I am a monster. But one carries people forever in one's self, in the blend—the secret blend of words exchanged, moments shared, silences, the experiences of colors and smells, all things human and endured together. But things, not.

A thing never becomes part of "the blend"—not in the same way.

At any rate not for me. A day will come in Manhattan when I shall miss the Sussex Downs, the cliffs, the green, the gulls. I shall miss England.

Glasgow,
April 1975

I am pretty fed up. Sometimes life is brilliant insight, and sometimes it is papers getting turned down again and again and again. At the moment I am very excited about going to America. It gives me a new chance. It is going to be like going home. The price of growing up can be terrible, yet a situation where no one knows you, no one cares about you, no one talks to you, no one interrupts you can be really creative. This is what it was like in London. And the price? Loneliness, of course.

I stayed in the lab until five in the morning. It was just me and the blackbirds. Actually, ecotaxis is being accepted—very gently now. I have real evidence too, but it is always exciting to have the thought without the evidence. Sometimes things happen so very naturally that it is difficult to remember when one was wrong. This trapping: Is it a spleen abnormality or a lymphocyte abnormality? Does it actually matter which it is? The question ultimately is: Why is there the depletion at all, why the trapping somewhere?

There is an inhibitory effect of brilliance, because what you do with small things is different.

We need Renaissance patrons of science. It is really terribly important to have them, and also the people that keep a lab sane.

April 24, 1975

You wrote of "the enterprise." The real contribution of "the enterprise" is not the one that begins and ends with just one mind. I could die tomorrow, or choose to stay in Glasgow tomorrow. The enterprise, the inevitability of collective thought, would go on. I truly do not think I matter. Hence my problems: I cannot fight for my own promotion or for titles, etc. It matters only that I should be happy. That matters, because I am more creative and more rich (scientifically) if I am happy and at peace. After all, the idea of lymphocyte maldistribution developed here in the quietness of my peripheral flat in a cold, gray city. But if I had not thought it, others would, because of the evidence which has accumulated in the direction of that thought.

What counts is the number of individuals in a circle of time, in a confined circle of space. But if the majority is mediocre, then the enterprise loses and it is a terrible loss. Because the mediocre will by definition choose the mediocre, and it will take time to break out of that circle.

June 15, 1975

In the midst of writing and rewriting my final M.R.C. report,* planning the theoretical—no, no longer that, actually, given the present evidence—paper on lymphocyte traffic and disease. Here I am. Examination month, examination time, serene and silent invigilating† time. Sitting in front of me is the best and most promising scientist of all our medical B.Sc. Honors students, writing for the most part a not very interesting exam paper. The back wall at which I am looking is, in fact, a large window secluded behind the dullest of curtains, with just a vertical slice of Glasgow stone to be seen from where I am sitting. The stone is the witness that we are just the protagonists. These reflections are springing out of invigilating time, and from the title of the review I'm writing for the publisher Chapman Hall, "Fact Versus Meaning in Lymphocyte Traffic." Lonely would be the echo without the sound; impossible the reflected image without the object; incomplete the fact without the meaning.

And somehow, as I sit here, do you know the feeling I suddenly get? The taste of British years, as if I've been a word in parentheses. Have you ever thought how words in parentheses must feel? I am sure it is good for a wild word to be put in parentheses from time to time. It learns self-discipline, discretion, humility, judgment, and by the time it finds itself out of them, it appreciates life, and the flow of language, and the stream to which it clearly belongs so much better.

As I sit invigilating, freely sketching the skeleton for the M.R.C. report, the reviews, and the paper, that is where I am now, out of parentheses. And I look back and see this huge, convex thick line, fading away in the distance, like station platforms metamorphosing into distant gray squares as the train departs.

*Her research had been supported by a grant from the Medical Research Council.
†Supervising examinations.

August 25, 1975
(tape recording)

The reason why I decided to come to the tape recorder today is because I may have made a little discovery, and I am extremely excited about it. I'll try to tell you what it is. I was checking the B-lymphocytes prepared for Sebastian from animals that had received repeated injections of ALS. This is an antilymphocytic serum that destroys the T-cells. However, this batch of serum was not working very well, for there were still a lot of live T-cells around. And what I found was a frequency of one T-cell together with one B-cell which seemed to be higher than I have ever seen before. I have done experiments in the past with these two cell types, trying to find out if in fact they ever touch each other. And they did so very infrequently. But today, in these animals with a slight depletion of T-cells, this event was much more frequent. Before I go away, I would like to repeat an experiment and compare the frequency of T-cell and B-cell pairs from a normal animal with those from an animal injected a couple of times with antilymphocytic serum. I will do extensive counts and measure the frequency with which the two cell types get together.

Now the reason I am extremely excited is that there is already a well-established phenomenon, the phenomenon of "help," that is very important for immunology: To make certain antibodies, T- and B-cells must work together. Yet if you reduce the pool of T-cells, you in fact get better antibody production against certain antigens than if you have a normal quota. Why?

Imagine that Sebastian's hypothesis is correct and that, with lots of T-cells around, factors are produced that reduce the adhesiveness of B-cells. Therefore, reducing the adhesiveness must also be reducing the capacity of Ts and Bs to interact, to produce antibodies. But suppose by reducing the pool of T-cells one is actually also reducing the release of the antiadhesive factors. Then one is actually increasing the chance of an interaction between B- and T-cells. So this would provide a basis for the mechanism of immunological help.

I haven't yet counted the cells because I didn't have the controls with normal cells, and without these there is no point in wasting my time. But the sheer excitement—I don't know that it can be put into very simple terms. If you go on one of these treasure-hunt outings, you follow all those arrows and things and eventually find the treasure. If you are a child, you just get enormous excitement from the hunt as well as the treasure. To me it is exactly the same sort of excitement. Has anyone tried to analyze it?

August 27, 1975

It is three o'clock in the morning. I started writing a National Institutes of Health grant application. It will flow. I don't even have to continue now. Both from that Lancet paper you saw in London, and from another paper I found in Cancer today it is clear that a defect in the membrane of all circulating cells (red as well as white) is perhaps sequestering them in the spleen and that—as the disease progresses—removal of the spleen is beneficial to the hematological system—the blood system—as well as the immunological system.

October 4, 1975

Just finished reading the review in the T.L.S., of Neruda's work. No one will ever know me completely without reading him. If I were a full-time poet of the windy cold day and the naked night, I would have wanted to be Neruda. There are few people in the world who make me rejoice for having been alive at the same time as them. Neruda is one. His poetry is formidable in its human dimension. He wrote in poetry all the things that should be the concern of every man and every woman.

Let me tell you about my week.

Busy, busy. I have finally come to terms with the reality of my immediate future and have worked hard. Having spent two or three full afternoons in the library, I reached Thursday and Friday in a state of great—too big a word—considerable depression.

The immunology journals have grown in all dimensions: width, length, thickness. There are some very good papers and a lot—a lot of detail. More and more being done in detail, just as a man looking at the borderline of the sea and sand might miss the tide flooding to the whole beach. And then me and my idea: Who cares? Why should I believe that it matters? Why do I have to believe that it matters? Why do I have to believe that I carry the responsibility of demonstrating and searching for the evidence of an idea, when I am sure hundreds of other people could do it, and if I died tomorrow it wouldn't matter tuppence?

I don't know if Gerald Edelman ever gets depressed, or if Jim Watson or Francis Crick ever gets depressed. But for the more ordinary of us, the strength that is needed to believe that what you believe in is worth pursuing is very great. And at the end of a Glasgow wet, windy week, it cries out. You have no idea how meaningful it is to have this project of ours going within myself. Meaningful in the sense that it gives meaning to all my moods, so I feel all that I am experiencing is not wasted. For people putting screws in cartons, people fitting electronic circuits, may eventually know that the supposedly creative scientist sometimes wishes she had a job like theirs. And I wonder if it wouldn't be much healthier to be trained as a whole man or a whole woman, and have to learn to make electronic circuits, or work in the field, or do something real as part of one's education, and then go back humbly to the responsibility of carrying an idea.

At the same time, I find my silent excitement about the distribution of lymphocytes growing. I thought for the first time of valuable, practical applications if it works. Patients with metastases [spreading cancer cells] have been found to have low numbers of circulating lymphocytes. It is just possible that this is so because of ecotaxis, i.e., the little things are now trapped in the metastases. If we manage to trace them we would have an early warning detective system of metastases distribution. Good?

Don't ever forget Stoppard.* First you are either a revolutionary or you are not, and if you are not, you might as well be an artist as anything else.†

October 9, 1975

It rains as if in the sky there was no other element but water. The day it did not rain, Glasgow was beautiful. Full quota of tired yellows and shouting reds.

The famous B.Sc. course in immunology started today with neither of the immunology professors here. One still unwell and the oth-

* Tom Stoppard, *Travesties*.
† Or a scientist!—J.G.

er lost somewhere in Europe. As the class coordinator, I said my little bit: Universities are places of learning, not places of teaching; in science when a fact becomes part of established knowledge, it is no longer part of the scientific process but becomes a tool, as solid as a centrifuge, without the fluidity that characterizes things unknown and questionable. Also, where the toilets were, I told them.

Then there was a meeting with Sebastian and the lymphocyte study groups. Altogether there were six of us, including Sebastian and me. Not bad for someone—Sebastian—who didn't believe in lymphocytes four years ago!

My discovery (see tape August twenty-fifth) is true and weekly reproducible. I am so excited about it that I am bursting. It doesn't allow me to sleep at night.

October 19, 1975

Eleven o'clock at night, so the university tower says. Parents warm, tired, and at rest in bed. They are taking the news of my departure well. My mother's eyes are stale with time and suffering, and with not understanding why people have to live apart from each other. Another group from our family is leaving, for Ecuador, away from workers taking over his business and people attacking him. And my mother suffers. The staleness of my mother's eyes makes me sad, and she doesn't know. My mother was young when I was born, so I remember her crisp dark-brown, shining loafy eyes, young as a child's.

October 20, 1975

There is a lot going on. The experiments on T/B frequency and pairing in animals partially depleted of T-cells are now in the eyes and hands of Wendy, who is repeating them critically and enlarging the bulk of solid information about the whole thing. I can't remember when I first put on tape the excitement of my first observation, either end of August or mid-August. Since then, it has been the hard repetition of questioning. I keep on understanding the devil's temptation to scientists every day. One has an observation and an idea and facts all round substantiating everything. And the temptation is to jump, to anticipate data, to publish, to tell, before the actual hard facts stand up like a Renoir horse cart that will survive time and winter to give shelter and peace in summer.

It is good to create and have the opportunity to look critically at one's own early days. Dr. Vera was a marvelous teacher in that way. There was never speed in publishing. Things waited and matured in our minds against the test of time, so by the time they were ready to be written, they were better, safer and surer.

November 1, 1975

Sebastian phoned at about half past eight, to ask me for coffee with a Polish visitor interested in metastases. So we had a marvelous and exciting evening, explaining to Anne, Sebastian's wife, that the substances he is talking about are, for the moment, "transparent." For even if we have them in bottles, the "factors" are more ideas than real substances. Basically, they are factors of recognition of neighbors. So if you are identical to me and I am a cell, I don't just push you away. But if you are a heart cell and I am a liver cell, we both produce factors—originally called morphogens—that will keep us apart. In this way you stay heart and I stay liver, and we don't get muddled.

Now, I always visualized embryonic cells making morphogens because they move apart and need to "settle." But by adult life, I would expect only the cells that travel "naturally" to make them, like those of the blood system, or tumor cells that start metastasizing. Thus a metastatic cancer cell could be the one that itself becomes changed and, becoming "foreign," is pushed away and departs because all the surrounding ones don't keep it in place. It is all a question of having concerned neighbors who keep an eye on you.

A cell can change, and normally this change is noticed and perhaps modified by its neighbors—the permanent ones who are next to it and the temporary ones, like lymphocytes, who visit it regularly. So everything is kept in a collective balance.

But suppose these neighbors are distracted by other events—looking the other way, as it were. Then a cell can change, and the change will be neither noticed nor modified. The extent of the cell's change could then be simply the result of some terrible disruption of the collective balance to which the cells normally belong.

This notion is, I'm sure, going to be most valuable in years to come because it can be easily tested experimentally. The beauty of it all is that such a simple concept extends well beyond the boundaries of minute biological concerns and goes wherever you want to take it, *like, say, survival of the fittest. Is the fittest really the fittest, or just the one that the others around have "allowed" to be the fittest?*

Immunology has been concerned with "temporary neighbors." Cell biology has been concerned with "permanent neighbors" in embryonic life, of little importance for "real" adult life. We have to bring them together. I have to do one other experiment to clinch the T/B interaction "message."

November 12, 1975

Fauré and very lovely results. Red carnations. Cold coffee. So I'm drawing the idea and the experiment.

We T- and B-lymphocytes travel most of the time in journeys of blood and lymph, and we rest from time to time: in the spleen, the lymph nodes, and Peyer's patches. We don't mix much in an obvious way, but naturally we talk to each other infinitely more than people who look at us realize. In times of great distress—such as when some of us are killed by antilymphocyte serum—the fact that normally we interact becomes more apparent. And when some of us, T-cells, are killed, a great increase in the number of pairs of B-cells and of B- and T-cells is observed.

That is basically it.

The "great" August discovery boils down to one page of spotted airmail paper, but be sure that the paper to be published will not look like this.* The implications are quite considerable, and the reality so beautifully simple, that one feels a little moved at the microscope.

The implications are that Sebastian's theory must be basically right, that from my work with him and Angelo our postulated control of lymphocyte traffic must have repercussions beyond ecotaxis into immunological function, etc.

And one wonders still what it is all about.

Later—date not known

I'm tired
of thinking,
of emotion,
of being overwhelmed by the fulfillment of predictions.

---

* Eventually published in *Nature*.

And I'm afraid that at the intense moment of final "polishing" of a discovery there is no compassion. It is all bloody "intellect," I assure you.

The compassion lies at the beginning, with the choice of subject and the question of relevance. But at this peak stage, all else but understanding is irrelevant. And perhaps it is wrong. I don't know. But it is impossible to stop it. This is the point of no return.

November 19, 1975

This morning, after spending the whole of last night reading, thinking, and putting the discussion of my own evidence on paper, I woke up terribly serious and with a sense of reverence I tried to analyze. I tried to think of moments before when I had felt similar reverence, for a thing or a person, and I knew exactly when and where they happened. At the end of a long walk in the Valo de los Caidos in Spain, where all, Reds and Francophiles, who fell during the Civil War are buried; in Rehovot, this summer, at the end of a long alley of trees that takes us to the burial place of Weizmann in the Weizmann Institute, where a sculpture to the dead during the Second World War is also placed. But I could not explain how, at the end of a journey of discovery when the more I read the more it makes sense and relates to what other people have found, I find myself so serious after such exciting moments of pure, unadulterated joy. I couldn't explain why my being at a point that now seems fundamental to the regulation of the immune response could make me experience this kind of reverence. How do we operate when we move into a deeper layer of understanding, of people as well as things?

In a way, our earlier finding of cells going here—or there—was a bit like that, but I knew it was not fundamental. For at that time we were not asking why they did that. I wondered if my knowledge of its lack of fundamentality was one of my reasons for not being too possessive about it, as I find myself now.

And my serious frame of mind, as I woke up this morning? One feels that one has touched something central to another person, or to a subject, and one feels silent and grateful in a sort of way, because one was allowed to penetrate a layer of understanding which remained impenetrable to others.

It is in a way ironical that I should go through all this well away from the glamour and wealth of New York science. The animal house situation is so bad here we cannot afford many animals. We are doing one experiment a week using a few mice. Today one experimental mouse died and we nearly gave him a State Mouse Funeral.

We may be poor but . . . ! Somebody gave a talk at lunchtime about the ampicillin rash in infectious mononucleosis. You have no idea what it is to go from the poverty of one experiment on a few mice each week to the enormous wealth of ideas, seeing the significance we are nevertheless deriving from other situations experimental and clinical, even while in that poverty. It seems that ecotaxopathy—a long name for the disease condition that follows from lymphocytes being stuck when they ought to be moving—could have the most serious repercussions in the depleted populations that get left behind. Not enough lymphocytes spells trouble. Such a concept can easily be of help in understanding lots of immunological paraphenomena—occurring during infection, for example—which at the moment are not the least understood, even in simple terms.

April 20, 1976

Somewhere, thousands of feet above the Canadian Arctic. All very unreal; all very, very unreal. I suppose reality will begin when I land. Today is just another airplane, admittedly with 180 kilograms of excess baggage, another heavy briefcase, another leaving the Glasgow mousery, another . . . like so many times. There is nothing to tell me, or make me feel, that behind are nine years—twelve, really—of Britain (thirty-six of Europe), and this time I shall not be coming back.

Who divided the land? Who divided us and made us feel black and white, Americans and British and Portuguese, when all we are are lonely primates wandering around in the wonder of the technology we initiated? Emigrating only once—to death—without passports. Who divided us and made us feel guilty when we leave? When all we belong to is our separate destinies. Even if for many that destiny is to share their lives with others. Who did all that? History?

East 81st Street, New York
April 30/May 1, midnight

Lots of practical things. Cleaning the floor yet again, and the oven, organizing the telephone, getting the first lab orders through;

meeting the finance coordinator of the institute, and the future technician; reading Anne Sayre's book on Rosalind Franklin; looking at the house and making it into a mousery within myself. Sitting still for hours and, like sand blown away by the wind of change, letting the past settle, strong and silent and deep, and letting the future begin. I've had little flashes of the beginning, with the re-encounter with the work left behind, and I know I could be in no better place than this to accomplish (accomplish is such a horrible word), to do, to ask, the relatively simple questions I ask within myself. But let us not kid ourselves. The longing is vaginal and naked. Saying, like Anne Sayre, that Rosalind Franklin's sacrifice was not to have children. Rubbish!

I don't know what a sacrifice is. Something deliberate and sacred? And life is neither. Life is this mysterious sequence of choices and changes, that finds you at thirty-six single, "associate member," and "good," depending upon the bureaucratic or friendly company you find yourself in. At heart one is none of these things, but the product of having said no to this and yes to that—not once, but two, three, perhaps four or five times.

Things went well at the core meeting. Things go well when other people trust you. It makes the "yes" choices easier.

The microscope is coming next week, and I am hoping to be ready to start.

May 7, 1976

Between these two occasions there was a site visit—an inquisitorial visit to those who are requesting a grant from those who may be granting the grant. It was also an occasion for self-doubts. Why does one do it? It is good to ask why. Why does one give up everything for something that is just a thought inside? I am having terrible doubts.

A lot more is happening than we could possibly have anticipated! My father's birthday is next Tuesday, and I bought him a book. A couple of people mentioned that the author refers to me, and that is the sort of thing that naturally would please my parents to see. They may not understand another word, but they sure can read their own name. So I bought it—only to be so embarrassed by the inaccuracy of the whole bit regarding me and the lymphocytes that I don't know whether I will send it to my parents after all!

For example, there are in pages 93 and 94 a series of references to someone called Coleman, that should have been Gowans. Much of

Chapter 8 is an echo of my own lecture in Minneapolis (the author is from Minneapolis), so that I suspect he picked up Coleman for Gowans from my "Scottish accent!" I don't know, but I am appalled (can't spell it). I really am. The thought of how a reputation can become public, clearly from two (rather good) lectures and virtually little else, is filling me with chills. How easy it is for a fool like Summerlin° to believe that if someone with an inaccurate mind says you are brilliant, you believe you are and then have to cheat to keep it up. "But for some people, self-knowledge is devastating," you once said to me. "C'est dommage," I reply. Much more devastating, in fact, is this kind of untruth. Who is this person to judge me brilliant? Who is this person who writes Coleman for Gowans? Who is this person who talks of us as if the whole world were one lecture theater in Minneapolis, and the whole of time two cold January days in Minnesota? Who is this person to call me anything?

But when I started by saying we have more than we anticipated, I meant that I never thought to be quoted so generously in a popular book, and have to share with you who are writing such a book the embarrassment of experiencing it.

Like the people from China who made the rush rug I now have in this room—with an accuracy, a passion, a beauty that matches any of our scientific days—I'm going back to my initial desire of being anonymous throughout your book.

I'm a little tired (12:42 A.M.), and would like still to do some work. The children's Hodgkin's disease stuff is just pouring out with very interesting results.

May 11, 1976

Yesterday there was the rehearsal for the coming site visit, this one reviewing the human leukemic program in which is included my request for $136,000 for three years to study B-cell lymphocytes. And we just went through the various groups here working in this field. Some superb, like the work of a young man named Edward that revolves around a central idea of the mechanism of the disease; other labs, though, where they are doing things, just doing things, in a cumulative—I would say accumulative—way. Terribly impressive in

* A dermatologist who painted the skin of experimental mice to make it appear that skin grafts had taken.

extent and sheer quantity, but not a fundamental thought putting it all together or apart. So I didn't sleep too well.

Why should I be asking for so much? Why should I believe that what I am proposing to do is of any relevance to human leukemia, or at least of any more relevance than what the chap next door is trying to do? Then there is this talk, that so-and-so who is coming as a site visitor, and who has not been funded himself, may be biased against us. And, of course, it is hard to expect such a site visitor to overcome his own humanity and saintly acknowledge that others are better than he. I feel enormous sympathy for site visitors.

Today I have no doubts about the probability of being right in my thinking and in doing the things I am doing. No doubts there. I believe that what I am doing is worth doing and may turn out very profitable in understanding the pathogenesis of numerous forms of lymphomas. I don't have any doubts there either. My doubts are about measuring that against the ideas or the work of the chap next door. So on Monday I shall stress the fact that I'm not doing anything that other people are already doing. What I am proposing to do is new. Also, from the analysis of the Hodgkin's disease study—Bless Tien-Chun, who is getting so excited about it all—it is clear that the results are turning out infinitely more valuable than I anticipated. You see, from the study of the children we are learning an awful lot that may turn out to be applicable to the treatment of adults.

And it is as if I have come around the whole circle, because I started in Glasgow persuading people to do thymus grafts in Hodgkin's disease. Of course, if I had thought about it, I would know that children are a natural model of thymus grafting. The results are just there, crystal clear, in graphs that anyone can see.

At another age, at another stage, I would be only excited. Right now, I cannot believe the evidence before my eyes, the evidence indicating that Hodgkin's disease may be a form of ecotaxopathy. For if you have cells (the lymphocytes) removed from circulation, and at the same time a center continually producing them in a compensatory way—as in the thymus of a young child*—of course the defect is more easily corrected and controlled than if, as in the adult when the thymus has stopped functioning, you don't have a source producing the cells as they are removed from circulation.

---

* Hodgkin's disease is more easily and successfully treated in children than in adults.

May 14, 1976

A rather important day. Preferential visa arrived; microscope arrived; first salary arrived; Dr. Bullock (leprosy man) gave lecture on the immunology of leprosy—all about experiments on the aspects of lymphocyte maldistribution I first talked about in Portugal in 1972. Exciting. Exciting to hear your thoughts coming out of somebody's mouth, as if we were all a great big thinking mass of different mouths in time and space. And because of the timing some mouths in space are heard, while others are not. But in the end it does not matter provided the collective brain continues to think.

May 30, 1976

Sunday: a quiet day, coming to its quiet close. Do you remember some biggish review thing I wrote and had ready on February 8? Too big, I thought. But my friend Francis, the editor that put me on to it, didn't think so. "Add something about cell traffic in embryogenesis, two paragraphs only, and surely more about the effects of the antigen, please," he wrote.

Now: just try to write two paragraphs and cover everything about cell traffic in embryogenesis! Never mind. It has been wonderful to be made to read some papers on the subject. It is all so simple in nature and, by God, so fantastic. Ecotaxis: how each cell right at the very beginning finds its way to well-defined microenvironments in the same embryo. The cleverness of the techniques is amazing, and what we can learn and integrate about all forms of cell traffic is really fascinating.

Spent all day struggling with those two paragraphs, plus a synthesis of the antigen effect. A day like this is a blessing, whatever that means in secular terms. A bridge of inner bricks between weeks and months. Solid bricks of the peace of enjoying immensely what one is doing, even if grumbling about those two paragraphs. And at the end of a first busy month here, it is good to have an absolutely quiet Sunday to reflect in English, the adopted language of one's inner thoughts.

July 8, 1976

Tell you about the beauty of science, you ask. Difficult. Unlike the stars and the trees, the object of our day-to-day doing-our-own-thing is not "beautiful," in the eighteenth- or nineteenth-century sense of the word. A leaf is beautiful in its own right. A star is beautiful in its own right. The beauty of a dead mouse and its labeled cells in suspension in a little bottle in a big gamma counter is, as I told you, the inner beauty of understanding, of listening to silence. To get the beauty of my science, the kind of science your cancer research scientists will be doing,* you have to like Stockhausen and read Cage. You have to learn beauty in the twentieth-century sense, and I think that in a way music has made the greatest progress in this respect.

But even I, a devoted Stockhausenite, and a Cageite, find it almost impossible to think with the continuing noise of the deep freeze in my present lab! There are, however, musical rhythms to discover in any lab where we count cells. My favorite beauties are, perhaps, in the radioisotope counters. But you have to like the beauty of the panel of instruments in a 747 cockpit to find beauty in a radioisotope counter. You have to understand the inner processes going on between the radioisotope and the scintillator to feel that you are in the presence of something really beautiful. Sorry to sound boring, but the best analogy is always love—making love between conventionally ugly people. The beauty is where no one can see, like the beauty of contemporary music. Let me quote Cage's Credo: "I believe that the use of noise to make music will continue and increase until we reach a music produced through the aid of electric instruments."

So a lab is basically a Cage symphony.

But one has to be fully aware of one's own culture, beyond the door with the radioactive sign on it. Not like a distinguished immunologist colleague in Holland, whom I asked, "Do you like Stockhausen?" to which he replied, "Can you get it here?" thinking perhaps it was a brand of sausage or some such German commodity. In a world of cells, the scanning electron microscope and what it enables us to see is the most dramatic expression of authentic beauty—even in the eighteenth-century sense.

* I was writing a film script on cancer research that summer.—J.G.

The visual experience closest to what is revealed by a scanning electron microscope? Seeing those flat black stones on the top of the mountain in Provence on that cold dark day. Remember? No color, just the beauty of form. Only the form of the cells is round. It does not rain or snow inside the gut, but erosion occurs; so sometimes at the top of the intestinal Alps, at the villi, the cells are flat because of diseases and disaster to their owner.

"Your silence, Cage, is an instruction to listen," wrote C. H. Waddington. Get this symposium if you can: Biology and the History of the Future, a UNESCO symposium with Cage, Heden, Mead, etc., presented by Waddington and published by the Edinburgh Press in 1972.

In some unexpected ways this may have been one of the most important weeks of my scientific life. A series of those Holy Ghosty coincidences, one after another, made me reach the hard wall of deciding not to have another thought (not to make public) for the next four years. It will be just a time of doing one experiment after another after another, until the results speak for themselves without my having to utter a word.

On Saturday I received a letter from Marion telling me that our paper "Ecotaxis: The Principle and Its Application to the Understanding of Mycosis Fungoides" had been turned down.

On Tuesday, walked to the very top of the hospital building, to the office of this young child, taking his administrative job with touching seriousness, and presented him with a new grant application for the NIH; he looked through its weight, the length of its methodological paradoxes—what a Freudian slip!—paragraphs, and in greatest embarrassment told me, "I hate to be the one to tell you, but this won't do. It just won't do. More detail, more detail, more detail." Then at some point during the beginning of the week, the rudest, the hardest, perhaps the rightest letter from Angelo, in reply to my comments about the discussion of his thesis; to crown it all, two letters from the American Cancer Society, one a flat turndown, the other grant approved but not funded.

What depth of conviction and history does one have in order to carry all this within oneself, to still stay in the lab until midnight and nine o'clock tonight? Where does the strength and resilience come from to face the flat turndown of the American Cancer Society? Actually, mine comes from one mouse.

My application was about its being possible, one day, to detect metastases by tracing the lymphocytes to sites unimaginable by other more conventional methods. And that one mouse, a nude mouse with retinoblastoma implanted in its eye and a thymus graft under its skin, showed me just that. For, most likely, the low lymphocyte count that I found in the blood is a reflection of the fact that the lymphocytes may be in the tumor. Strength is also being derived from seeing one kid—a volunteer student—growing in delight and concern for the results of her little experiment, reasserting the sense that discovery is a worthwhile human activity. Discovery of the simplest nature, like finding a goldfish in a remote Moorish well.

So from now on, it will be just a tight, narrow road between narrow experimental rails, a situation fit, still fit, to my eye and status. I long to be old, and resent deeply the fact that your ideas may not be worth an article, or a penny, before you are fifty, and even then, who knows?

There are enough turned-down manuscripts in my filing cabinet to keep my creativity under the tightest experimental control. The irony of it all is that at the same time I must become a more selfish being. One doesn't want to make public one's ideas at the early moment of their conception, before the evidence, or the experiment, for fear of being unable to do the experiment. But at the same time one does feel an enormous desire to share and let other people multiply the experiments that will prove them wrong or right. If people looked at the terribly worried faces of the visitors in the hospital halls, people visiting patients, they would get the feeling of urgency. There is urgency in resolving biological problems that are causing people to suffer, as there is urgency in changing societies where police can kill innocent people expressing their overdue protest.

You are lucky to be a historian, and I am lucky to have well within myself some remote and deep sense of history. Virtually no matter what happens to me as your subject, like an interesting case to a medical scientist—no matter what happens—it is of material interest to you. I am lucky to have the sense of history because it helps me to hold the otherwise untenable view that I'm right, or at least that I can be right about this, and everybody else in the world wrong, in this sense: that the difference between right and wrong are not differences of weighable, long, heavy grant applications but differences of *that most important of all dimensions—time. That in a few years I shall either be proved right or wrong, and either event does not deserve my present . . . what? I don't even know what word to put in there. What I feel is tired. And you know of what? Of being young.*

July 17, 1976

Here my first wonderful scientific thing happened, like a balloon, like its likeness and colorfulness and its wonder to a child. Dr. B. wants me to collaborate with him and his group. The very experience of sitting in his room, waiting for thoughts and words to be articulated with all the measured hesitation of brilliant determined Englishness and almost uncontrollable creativeness is something that has put me in a state of very primitive delight. The same delight I first experienced when I went to London and listened to Dr. Vera, and the same delight as thinking with Sebastian. To have to worry about money, grants, deadlines is an awful thing.

Being here is starting to dawn on me as the challenge it is. The wave of problems to be tackled is starting to swamp me, and it is a marvelous feeling. I was going to say that I am walking a bridge between nostalgia and challenge and went to the dictionary to check the meaning of challenge—"called to account." "Who goes there?" Yes, "challenge" is right. Walking on this bridge is like moving between two countries on one of those long bridges over the Danube or the Rhine. But slowly the balance is moving toward the challenge, and the nostalgia is incorporated into the challenge itself.

July 21, 1976

I can hardly describe the experience of spending those "thinking hours" with Dr. B. It was the liberating experience of walking in Sussex on the Downs: anywhere of immense space and green, where eventually you will get somewhere, not because the route is posted for everyone to see but because you know you are going somewhere. With my friend at Harvard, working on microcirculation, it is different, like picking up pebbles. One goes down in detail and it is marvelously refreshing; and one knows, too, that one will get somewhere, not by moving but by looking at what is already there to be seen. Cancer is an urgent problem, but no matter how much money you put along the posted route, it will only be solved—and I mean solved—by those who wander. Meanwhile, I am working seriously on a leprosy project, and it has been quite an experience. Tomorrow will have to go to Rye to kill leprosy mice and, therefore, get up at six o'clock.

CHAPTER 7

A DIVERSION
AND A FAILURE

Defeat is a school in which truth
always grows strong.
HENRY WARD BEECHER, Proverbs from Plymouth Pulpit (1887)

Though in the Bible it is referred to as Lazarus's disease, Lazarus possibly didn't have leprosy at all, for his clinical symptoms are not recognizable, and in any case, in earlier times, the word "leprosy" was applied to a whole range of chronic skin conditions in which the skin becomes scaly. It has also been called Hansen's disease, after the man who, in 1874, was the first person to see the incriminating organism under the microscope. That event was a double first: Not only did Hansen identify the organism responsible for leprosy, but that bacillus was also the very first organism to be identified as a causal agent in human disease.

Leprosy is a chronic condition which affects the skin, the eyes, the mucous membranes, and sometimes the testes and nerves. It is only mildly communicable and thus never deserved the revulsion and the enforced isolation with which it and its victims were treated until recently.

It is believed that leprosy was introduced into Britain by the Roman legionnaires; it had disappeared by 1798. But even in 1960 it was endemic in certain parts of the European continent and in Central and South America. Ten percent of the population in Africa still have it, and so, too, it is estimated, do some nine to twelve million people throughout the whole world. It seems to flourish in damp, tropical climates; even in the United States there have been small outbreaks in the Gulf region and in Southern California.

The organism responsible is *Mycobacterium leprae*. The word comes from the Greek *myces*, meaning a fungus, and the characteristic branching of these bacilli resembles the growth of fungal threads. The bacillus is one of a group that in growth, appearance, and staining—the laboratory procedure which colors cells so that their structures stand out—has certain definite characteristics. The bacillus that causes tuberculosis belongs in this group too. They are all classified as acid-fast organisms because they absorb certain acid dyes. In an infected body and in favored places in and among the cells, the bacilli form *globi*, bundles of microscopically small sticks. But though leprosy bacillus was indeed the first to have been identified as an agent of human disease, ironically and tragically it has turned out to be one of the most refractory, refusing to be cultured in laboratory dishes. Thus, because of our inability to grow the bacillus in the laboratory for a sustained period of time, even in small quantities, let alone the vast ones which would open the way for a vaccine, our efforts to prevent the occurrence of leprosy are likely to remain stymied. The best we can do is look hard at human leprosy in human cells and use mice and rats infected with rodent leprosy for an experimental model of the disease. We struggle to understand as much as we can this way, while still trying to persuade the human organism to grow in our laboratory cultures.

When Anna first came to America in 1974, on her six-month sabbatical leave, a small study with the animal model was already under way, initiated by the director of the institute. Like many other immunologists he wanted to know whether there is a genetic susceptibility, or resistance, to certain diseases; and, if so, whether this susceptibility is linked to a particular portion of our chromosomes. Immunologists have found that our capacity to reject or accept a graft and to respond to certain foreign substances seems to be governed by a group of genes at one particular point, known as the *histocompatibility locus*. Resistance or susceptibility to certain diseases is turning out to be similarly governed.

The director had initiated the study because he wanted to know if a similar genetic property was at work in the case of those leprosy patients who do not respond to conventional therapy. So a laboratory model had been set up by a visiting Korean dermatologist, with the aim of trying to mirror in rodents the problem in patients.

Anna, like all scientists, seizes a good opportunity when she sees one. Because in leprosy, like Hodgkin's disease, the absence of lymphocytes is a recognized diagnostic factor, she had a direct interest in the model, seeing in it a chance to study lymphocyte traffic.

To get the model going, mice are first injected with the organism of rodent leprosy, *Mycobacterium lepraemurium*. These invade the spleen and the liver, where they settle down to divide. They do so in very precise sites, and a very well-developed infection results. By sectioning the organs and then staining the sections and examining them under the microscope, the invasion of the animal by the bacilli can be traced. In the early summer of 1976, Anna herself started to look at some of the sections and so came upon a new, unsuspected aspect of the problem that was to involve her and a number of scientists for more than three years.

Looking at the slides, she was immediately worried that the mice had acquired an additional infection, over and above the intended one of leprosy—worried to the extent that she felt she needed the help of an expert microbiologist who could look at the slides and say confidently that the mice were full of the leprosy but absolutely nothing else. At the same time, the director became interested in trying to initiate the growth of the microorganism in vitro, in the laboratory culture dishes. For years many people had regularly attempted this, and outside reports suggested slight indications of some possible progress. Therefore, in response to both their needs, Marie, a microbiologist from Indiana University, arrived to sort out the microbiological side of the project.

The leprosy bacillus actually grows in the macrophages, the "big eaters," those white blood cells whose function is primarily one of scavenging. Those macrophages that scavenge in the bloodstream are called *monocytes*. By literally gobbling up all invaders, they help keep the body clean and free from infection. But we tend to forget that these sentinel cells are just as prone to disease and infection as those they protect, and in leprosy the very cell that most needs to be healthy and flourishing, the scavenging macrophage, is in fact the site for the infecting invasion. In the laboratories in Rye were a whole series of macrophage cultures that had already been maintained for long periods. The new idea was to infect these with mycobacteria and see what might happen. So in the summer of 1976, under Anna's general supervision, the new series of experiments started.

Meantime a new arrival had appeared. Nara is an Indian doctor and a pathologist at one of the main hospitals in Rhode Island. Because he and his superior felt it would be useful if he had some experience in immunological research, he joined the institute for a year and agreed to become associated with the leprosy project. Anna, wanting to make him feel at home, gave him the slides she had been studying and asked him whether, as a pathologist, he saw the same things in the slides that she had seen. But before very long Nara was feeling frustrated. He had really come to do immunology, and here was this project, getting more microbiological and less pathological every day. "I'm not learning anything," he told Anna. And then there was another problem. "Nothing happens. No one tells me what to do." But she had quite deliberately *not* told him, and she now explained that he would have to find his own scientific question, one that intrigued him, and then work out his own experiments. She would, of course, help him in every way possible, but, she repeated, she was not going to *tell* him what to do.

This is a recurring problem when doctors come into scientific laboratories for the first time. A doctor doesn't have to initiate anything because something is always happening. Either people die or they don't. In either event they are first ill, and what needs to be done is frequently quite obvious. Even a pathologist in a hospital doesn't need to precipitate action, for material keeps coming in to be studied and judged. But in a scientific laboratory nothing startling which calls for an obvious response occurs on its own. Nature is passive and silent. Scientists have to start things up for themselves.

In early August, Anna went off on vacation. Before she left, she suggested to the others that they add some lymph to the system of cultured macrophages and bacilli, feeling this might make a difference. Why did she think the lymph might help? Because when she first started to examine the sections of infected mice, she saw something intriguing. The infection develops in the macrophages very early, especially in those of the thymus-dependent areas of the spleen and in similarly specific areas of the liver—in fact, in areas associated with the lymphatic routes. She knew one important thing about these areas: They are the places where the lymphocytes go into the lymph. For a long time she had been puzzling over one problem: Why should lymph be such an ideal medium for tumor growth and spread? For it is in the lymph that metastatic cancer cells thrive and are carried to other sites in the body. The fact that mycobacteria also thrive in these particular sites suggested that they might also provide an advantageous environment for mycobacteria to divide and multiply. Adding lymph to the cultures seemed a logical step.

At first her colleagues didn't take her suggestion seriously and didn't add the lymph. But finally Marie started a new culture series, and added some bovine lymph to the cultures containing macrophages only. It was not added to the cultures containing macrophages and bacilli. The only result was that in high concentrations the lymph was slightly toxic for the macrophages. At low concentrations they seemed indifferent to it. When Marie had to return to Indiana University, Nara just kept the system ticking over slowly, slowly because he wasn't interested in anything microbiological. Finally, either out of boredom, or desperation, or because on her return Anna gently reminded him, he added lymph to the cultures of macrophages infected with the bacilli of *Mycobacteria lepraemurium*.

This was in October 1976. Three days later the culture was packed full of the mycobacteria of murine leprosy. He couldn't believe his eyes! And when he finally told Anna he had something to show her, there was a culture system chock-full of mycobacteria that had been kept going for about seventeen days! But because the experiment had been done without controls, at that stage it wasn't scientifically "proper," and Anna said nothing to anyone. The following week, Nara repeated the experiment with controls—that is some cultures contained added lymph and others did not—and in the lymph-enriched cultures the mycobacteria not only seemed to be thriving but their numbers seemed significantly higher.

When I first asked Anna why she had not told the director about the first results, she said, "I avoided telling him too early because he might get excited. One of the reasons why there is no vaccine against leprosy is because nobody has managed to grow it in the laboratory. There are lots of little techniques here and lots of little techniques there, but no one has yet unequivocally grown the stuff in sufficient amounts to do anything with it. If we had a method which allowed the rapid growth of this microorganism in a culture dish, it would be unbelievably exciting. So I had to be sufficiently sure that there was something there—genuinely there—worth speaking about." When in late November she was entirely certain, she told the director. As expected, he was intensely excited.

On December 3, he was due in Anna's laboratory to look at

stained smears from the crucial cultures. As Nara was getting the slides ready, he told me how very appropriate this occasion might prove—just 102 years after Hansen's first identification of the bacillus. He told me too how fussy this type of bacteria is, infecting particular animals only, and having a predilection for invading certain organs. Tuberculosis bacillus is equally fussy.

Anna arrived and began to examine the batches of slides. There were two sets: one from the first batch of experiments without controls, the second showing the experiment repeated with controls. It was a nice moment. When the prepared slides were sharp and clear—which they were not always—one could see the little packets of sticks, small cylinders stained bright blue. Anna was quiet and a little white. She spoke only enough to make certain that she knew precisely what she was looking at in the time sequence of the experiments, ordering the necessary data in her mind so that she could present them properly. Otherwise she was silent.

But no one else turned up. There had been some mix-up about the time. I asked Anna about her white silence, and she told me she had been silent because she was angry; and she was angry because the second batch of slides—the important batch—had been quickly and inexpertly stained. They were, in fact, a horrible mess. She had been forced to keep quiet for fear of letting her anger show.

She explained, "They were very important slides and they were messed up, so the natural reaction was anger. But you must *never* be angry, because if you are angry the next thing they are going to do is to hide the results from you, like children, and the most natural person to hide the bad results from would be the director. Or they are going to give you only the results you want. So you must never show your anger."

When the meeting finally did take place, Anna made Nara take all the slides along, including the bad ones, having first explained to him one fundamental difference between science and medicine. In medicine if you make a mistake, people die. So it is very important that you do not make mistakes. In science, however, it is very important that you *do* make mistakes and, even more, that you are not ashamed or worried about them. Only then can you progress.

"I made him take all the slides, the good, bad, and indifferent, and I said that not taking things—not acknowledging mistakes—is what other people in the past have done, and they got themselves

into a real mess." And, as it turned out, the director vigorously reinforced her statement, saying exactly what he should have said: that
it didn't matter; these things happened; the fact is so fundamental
in science that those who come from medicine must learn the lesson
immediately.

Actually, the meeting was unalloyed joy, with strong overtones
of farce. On the bench was a double microscope arranged so that
two people could look down it at the same time. The director took
up position at one eyepiece with Nara at the other, feeding in the
slides in sequence. As Nara showed the first slides from the first experiments, the director, in enormous delight at the sight, said what
his old scientific boss might have said: These things are best described in musical terms; it was musical, like the great chorus from
The Messiah. So there was the director of the institute, going off
into the future with enthusiasm and trumpets, talking about the essence of discovery and declaring that if you think you've made one,
you have to drop everything and run with it. Take Jenner, for example: After he had discovered that milkmaids generally did not
contract smallpox, he felt this same deep imperative. So he left his
family practice, dropped everything to go straight after his discovery, and stayed with it until smallpox vaccination had been nailed
down. Now Nara must go after this one.

But Nara was shaking his head in utter dismay, not really having
his heart in microbiology and cell cultures, not wanting to follow up
this discovery at all, searching around for reasons to let the mantle
of Jenner, please, fall on someone else—Anna, for a start, for she
had said to put in the lymph. Besides, this was not the assignment
that his chief had given him. But the voice at the other side of the
microscope was relentless: His chief would surely appreciate a discovery and where it must lead, and surely appreciate its value, surely understand why Nara would have to go after it and be really responsible for it. Totally unconvinced, and facing the dismaying
prospect of leaving pathology, Nara said, in gentle despair, "Oh,
please, don't force me to make a discovery!"

Finally, the director went off in search of the president of the
institute, himself a microbiologist, for Anna had insisted that at this
stage she would show the material only to a microbiologist, because
if the phenomenon was not genuine, it would be easy for them all to
seem like fools. When they were alone, Nara began again.

"I don't want it. I'll stand on the sidelines and applaud. But all I want. . . I'm tired. I've been a student for too long. I don't want to learn microbiology. I want to do immunology, look at simple things, and do my pathology, and go on in my own quiet way."

At which point the director returned, having failed to find anyone, and to Anna's obvious relief the trumpets faded away and the meeting proceeded in an atmosphere of cautious reticence. The director himself became skeptical. The growth had been so rapid that it could have been typical of other acid-fast positive bacilli, like tuberculosis; they might have thought they were looking at leprosy, but it could easily have been some other infection. As soon as *that* thought struck him, he began to doubt seriously; he became, in fact, extremely skeptical. There had been too many variables; it would all have to be repeated, as Anna had insisted, in the microbiology laboratories, without any possible contamination or infection. But later that night Anna said to me that the best bit of all was to see him "smell," almost like a dog, the potential of a discovery, because he had the capacity to see the significance of it.

"What," I asked, "apart from the vaccine, is the significance?"

Anna was horrified. "Apart from the vaccine? What do you mean? Do you know how many millions of people are affected by leprosy? There is no need for it to be significant apart from that. But, of course, he can also see the significance of the technique for other bacteria, and other diseases, and he can also see the significance for cancer spreading, as I do. These mycobacteria utilize the lymph for growth, and tumor cells utilize it for spreading, and *that* is why I am interested."

"I can see that if you can do it for one type of mycobacterium you can possibly do it for all similar diseases. But what refractory ones are left?"

"Fungal diseases, for a start," she replied. "A number of classic bacterial infections besides; and a series of odd bacterial infections that seem to turn up in immunodeficient patients about which nothing has been done. And if this is true—if through the principles like this, of bacteria growing in lymph regions—if this is true. . . I am sorry to be boring with all these 'ifs,' but I am very reluctant to get excited, or let anyone else get excited. The day the microbiologists in their own separate hands do it, then I'll be excited. I really am sorry to be so tiresome but I find this very necessary. Because if I get excited then there is no one controlling the excitement, and that could be terrible. But if it *is* true, then it could be of tremendous significance in microbiology and for me too. It would be very significant for certain forms of cancer, where you have the tumor spreading along the lymphatic route. The conditions in the lymph are *totally* different from conditions in the blood serum. There is a genuine molecular basis for this fact. But once you have got hold of a principle—which we might now have—that applies to one situation, it could apply to others."

"You mean," I asked, "that if the lymphatics provide a fertile environment for the growth of mycobacteria on the one hand, and for the spread and growth of tumor cells on the other, then there might be something common to both situations—things in the lymph that both tumors and bacteria need? So you could possibly culture tumor cells in lymph, too, and study them with more precision than you are able to do now? Or perhaps you could modify the patient's lymph so that it wouldn't provide such a comfortable environment for the cancer to grow?"

"Right."

"Where do you go from here?" I asked.

"Well," said Anna, "everyone is going to start again. At the end of seventeen days, Nara's original culture just stopped growing, or the cells just stopped dividing. So he doesn't have that culture any more. We are going to start all over again."

"Will it run through your lab?" I asked.

"It can't," said Anna. "I don't have a microbiological laboratory. This is no kid stuff. This is very serious. It has to be run in a microbiological laboratory by people who are competent. I am willing and quite determined, if I can, to do some of the work. What I can do in my laboratory is count the things under the microscope. I can do that and I intend to do that, with double-blind samples. Because this has to be done blind as well. People must not know what the samples are. Talk to me in two weeks' time, and we will see if I'm going to be excited. Meantime, we have to have another meeting."

A second meeting had been arranged for a full-scale presentation of Nara's slides to the head of the Division of Microbiology at the institute, a division that has had vast experience with a whole range of infections. On that occasion, five people were present—the four scientists and I, the outsider. First of all the director filled in something of the background: how the initial work had begun because of an interest in refractory leprosy in human beings; how, with the help of their colleague from Korea, they had tried to set up an animal experimental model; how, after Anna arrived, the microbiology was straightened out, with Marie's help. He mentioned two recent claims, one reported in the press, of success in culturing the leprosy bacillus in the armadillo. But he was skeptical; nobody thought the work would stand up. He assured everyone that not a breath about the present work would reach any person—let alone a newspaper—until it had been totally, unequivocally demonstrated again and again, and by people other than those who had made the first observations. He then explained that through Anna's interest in cell traffic, she had looked at the lymphoid organs in leprosy and had noticed that the mycobacteria in the diseased animal were located at points where there was a heavy lymph flow. So she had said, "Dump in the lymph."

Nara then took over and described the sequence of events in the crucial culture where, on the third day, he saw pellicles floating at the top—groups of cells in which the number of bacteria seemed to have trebled. He handed out the Xeroxed notes which gave all the numerical data, going right back to those cultures started in August, and then began to speak, giving the fullest details of the experimental procedures. A veritable cornucopia of technical information poured forth. Finally we came to the critical culture, number J774 (20) A.

On November 2, 1976, infected cells were taken out of a lymph node dissected from the foreleg of an infected mouse. A culture was started but no lymph was added, and on the third day bacilli were seen in large numbers. They persisted through day 10 to day 28, by which time there were no viable cells left. Smears had been made and the bacilli would be counted. But on November 9 the next subculture was started—the crucial J774 (20) A—with cells taken from the same source and from the same genetic strain of mice, but now a 5-percent solution of bovine lymph was added. Three days later a pellicle had formed at the top of the culture flask, and smears from this showed "overwhelming" numbers of the same acid-fast, positive organism. One week later there were plump, fat cells in the culture gravid with the infection. But again, by day 24, the culture tapered off, and once more there were no viable cells left. On

November 26, at Anna's insistence, they started the experiments once more, repeating with the control cultures.

The pitfalls were discussed yet again. Had they merely uncovered the most likely of organisms, a *mycoplasm,* a fungal infection of the smallest, most ubiquitous, most irritating kind, the curse of scientists who use tissue-culture techniques? If they had uncovered something significant, would whatever it was grow in other media, or in other cells besides those mouse macrophages to which lymph had been added? If the process was genuine, what was the lymph doing? Was there perhaps some other type of acid-fast organism inside and outside the cells, or was there *really* an increase in the numbers of the mycobacteria?

The head of microbiology was then asked if his department, using their best technicians, would see, first, if they could reproduce the results at all. "You set the priorities for the rerun," the director said, and he went through the procedure, though every scientist in the room could have recited it by heart. What do we actually have here? When we know that, can we make it grow? If we think we can make it grow, then it's time to do the Koch's postulates—that is, to take the mycobacteria we ourselves have "grown" in vast numbers—we hope—and reinfect a mouse to see if it comes down with leprosy. If at that point it has worked all the way through, he said, the fun is over.

Of course, if the technique *did* work, they would next take cells from one of the *human* patients to see if they could grow the *human* mycobacterium in culture also. But in this case the Koch's postulates could not be carried out. You cannot take the leprosy bacillus from a human patient, culture it to the extent you want, and then infect a healthy human being in order to find out if you really did grow the bacillus of human leprosy. All you could do, if you thought you had been successful, would be to try to infect healthy human cultures and see if the disease declared itself there. Everyone should move fast, the director concluded, but circumspectly. Get the work cleanly and clearly done, repeated . . . and then published.

So it was agreed that Nara would take one of his newly infected mice to the microbiological department, where the infected lymph node could be dissected out under the most sterile conditions. One of the department's most competent technicians would repeat the whole procedure, possibly with a few extra flourishes of her own. A few more technical details were agreed on, a few more technical questions answered. A blow-by-blow account was elicited, not only of their culturing procedures but of their methods of counting the cells and the bacilli. From time to time, the director was singing in counterpoint, reflecting on BCG, our vaccine against tuberculosis, which, though it is derived from other animals, works well in human beings. Wouldn't it be fantastic if one could use the mouse strain of leprosy for humans? Anna shook her head cautiously but said later, "Actually, for a couple of amateurs, we haven't done too badly."

Two days later I met Nara and one very sick mouse. Like a condemned prisoner, it had spent its last night in cold, sterile surroundings. As we went over to the lab, Nara talked about the odd predilection of all mycobacteria to infect certain sites. Some bacteria always go into the myocardial lining of the heart, others into the sheaths around the nerves. One would really have to understand that cellular environment very precisely, or so he thought, in order to set up the right conditions for the growth of such an organism in a liquid medium rather than in macrophages. Unlike some microbiologists, both he and Anna were convinced that these ideal conditions would never be ascertained by simple empirical trial and error alone.

In their new modern quarters, the microbiologists have a charming reception area, with a notice over a box saying, "Please leave specimens here." The laboratory activity goes on behind partitions, unseen and unsuspected by the visitor. But entering their old laboratories was like going into Grand Central Station at rush hour. It took some time to find Ann, the technician. She sat Nara down in front of a laminar-flow bench, covered by a hood that pulls clean air over the surface and away from the experimenter. He asked for sterile gloves and disinfectant, then reached his hand in for the mouse.

"Wait a minute," said Ann. "I'll get you a corkboard."

Nara drew the live mouse out of the bag. The few faces staring at him were a study—a mixture of horror, superciliousness, incredulity, disdain, and a desire to get away. The odds were that never before had a mouse, dead or alive, appeared in that room. Microbiologists deal with living specimens certainly, but these come clearly and sanitarily in little tubes, bottles, or flasks. Neither patients nor mice enter their lab. Nara and I were laughing.

He sat down and picked up the mouse. I could see the enlargement in the pit of the left foreleg, the mass of the infected lymph node, swarming with leprosy mycobacteria. As he worked away, Nara spoke once more about the celebrations of Hansen's first discovery and about pathology, which he loved, with its own quiet routine. Then, casually, he broke the animal's neck.

Next, Nara picked up the detergent spray, sprayed the skin, and then laid the mouse on the board, all four legs akimbo. He unwound his packet of sterile instruments and with alcohol swabs cleaned the area. With scissors he cut the skin and with a scalpel he dissected out the swollen lymph node and popped it very carefully into a little sterile container. The lid was put on and the container handed to Ann. From then on this node would be her problem. She would establish two separate cell lines, one uninfected, the other infected with mycobacteria from the lymph node. She would use for the cell's growth a tissue-culture medium containing fetal calf serum for extra nutrition and penicillin for health, and some of the cultures would have lymph in them, too.

Because of the potential in the situation, she would start straightaway. This was most unusual. In large research institutes there are many people who not only can help you but whose help you need. Tissue cultures must come from one department; electron-micrograph photographs of your cells from another; slides, graphs, or diagrams from a third. But you have to be patient because other departments have their own jobs to do, and so it takes a long time before you know whether a critical experiment has worked or whether a result vital to your progress has any answer that is at all interesting. But in this case, the routine would be started immediately, even though the results would take a month to judge.

The month passed, and Ann's results were the same as Nara's. On the third day there was a real flare-up of the bacteria cultured with lymph, and the smears showed that the cells were heavily loaded with the bacilli. But within twenty-four days, the cell line died.

The next person to try was Mrs. Huang, the most experienced senior technician of the whole department. She undertook to repeat the procedure, to get her own quantitative estimate of the multiplication of the bacteria—if it still occurred; to see whether or not there was anything she could do to maintain the culture and keep the whole process going longer. One can never honestly say that one has grown an organism in culture unless, in addition to all the back and forths of infection and reinfection between cells and mice, one has both a real measurement of increase in numbers and has managed to transfer the infection through a whole series, from one set of cells to another set of new cells to another set of new cells—reinfecting each time.

It wasn't until a year later that I received from Mrs. Huang a full account of what she had done and what had been the inherent difficulties. She had made not two cultures but a whole series—with controls—using various dilutions of infected mice cells in suspension. She added these not to macrophages but to the cells usually used for culture, human fibroblasts. (Anna was not at all happy with this variation.) At some dilutions Nara's initial results had been duplicated, and Mrs. Huang had seen a peak in the numbers of mycobacteria around the eighth day. But she was not prepared to say that there was anything conclusive here. By the sixteenth day the cells had run their living course and the activity had completely stopped. The problem was that every fourth day she had to split the cultures; otherwise the cells got overcrowded and died. She tried again, this time using Nara's macrophage cultures, which she divided into two. In one batch she estimated the number of cells and then gave them to Anna, who counted the number of bacilli per cell. With the second batch she reinoculated a fresh culture flask and ran another whole series. There was still a difference between the cultures containing lymph and those without, and again only those in certain dilutions showed real activity, and again the division consistently peaked on the eighth day. Between day 23 and day 35, she was still counting, but the culture cells were dying. What, I asked her, were the basic problems in this whole business?

There were many. After a time there were many cells in the cultures, but not so many mycobacteria. One reason was that the culture cells grew so fast that they came floating up, crammed full of bacteria. Then, packed with infection, they died. These dead cells had to be removed because, like dead fish in an aquarium, they foul the environment so that the remaining cells can't breathe, and this marks the time when the cultures *must* be split. The hope had been that the bacteria in the dead cells would infect the new ones created by cell division, but they didn't. They wouldn't even come out of the dead cells.

There was the question of timing, too. If you want to grow infective organisms in a culture using living cells, it helps enormously if division of the cells and the organisms are in phase. But more often than not, they aren't. The culture cells divide very fast, and ideally the infective organism would keep pace. But mycobacteria divide very slowly. Consequently Mrs. Huang felt that to go on using cells was futile; growing the leprosy bacilli would work only if you could find the right culture medium and thus dispense with cells—whether macrophages or fibroblasts—altogether. The operative word was "right."

"You see," she explained, "the same general problems always recur: the problem of controlling the cell division in phase; the problem of the bacteria not passing from cell to cell. Maybe using fibroblasts we don't have the right cells. The mycobacteria certainly do grow well in the mouse macrophages, and they certainly did flourish and multiply at the beginning."

Maybe, she was prepared to admit, it *could* be done; maybe one could take cells full of bacteria and break them up by ultrasound waves, get the bacteria out that way, and so infect fresh cells. But that cellular route was not the route she would follow if she wanted to solve this problem. She would look to a synthetic medium. Initially there had been the same problems with the tuberculosis bacillus, but ultimately microbiologists had been able to duplicate enough of their microenvironment to grow them in a liquid medium. Mrs. Huang would thus start with the medium that worked successfully with tuberculosis and then go on adding various substances until she hit on the right combination: a thoroughly empirical trial-and-error approach which others had tried before.

As for other problems, Mrs. Huang had had none with contamination but only with timing and temperature. A temperature of 37 degrees centigrade was ideal for the *cells;* above that they divided so fast that they overgrew the bacilli. Dropping the temperature four degrees to slow down their division made life a little too chilly and the division too slow; at 35 degrees, the cells started up growth and division. The addition of the lymph did make a difference, she felt, but only about 0.1 or 0.5 bacillus per cell. Anna and Nara felt it was rather higher.

There was not enough firm proof to merit publishing the results, and in the event no one from the group published at all. But another group did. In September 1977, a Ph.D. student doing part of her research with Anna brought a paper into the laboratory which had been completed that April and published through the American Society for Microbiology.* The work it reported must, therefore, have been under way at the same time as Anna and Nara's experiments. The two authors of the paper, working at the National Institutes of Health, *were* growing the mouse leprosy bacillus in macrophages; they did not use lymph; but they, too, claimed a marked multiplication of the bacillus in the cells and interpreted this as an increase in numbers. By then, however, it had become clear that there was a real problem in counting; the bacilli get redistributed all over the culture, and some are taken up again by large cells. So as with body counts in wartime, it is only too easy for the same bacillus to be counted two or three times over.

Even so, Anna was confident of being able to make a genuine dent in the problem. During 1977 she had become more and more aware of the extent to which bacteria utilize or need iron. She therefore surmized that iron, or iron-binding proteins, might help the growth of the mycobacteria in the macrophages. Some fifteen months after Nara had excited us all, she discovered to her great irritation that microbiologists had known about iron and bacteria all the time; that in the case of tuberculosis, for example, the mycobacteria grow much better in an iron medium. In fact, the mechanism for *killing* the mycobacteria of tuberculosis is to give them *transferrin*, a protein which binds to the iron, thus depriving the mycobacteria of it and thereby "starving" them to death. "They should have told me this," she said. "We might have done something."

At any rate, in the summer of '77 Marie came back for another effort. This time iron goodies—transferrin and veal liver—went into the cultures. In one group of cultures, the number of mouse leprosy bacilli zoomed off beyond counting; in yet another it shot up higher

*Akira Yamagami and Yao T. Chang, "Growth of *Mycobacterium lepraemurium* in Cultures of Macrophages Obtained from Various Sources," *Infection and Immunity* 17, No. 3 (1977): 531.

but not tremendously. Then, as before, the cultures faded after some weeks. This again led to doubts about the counting techniques. Anna herself counted the bacteria by first counting the macrophages that held them. But was the added iron merely helping more macrophages to *eat* more mycobacteria and was that why she saw more and counted more? Or were the mycobacteria actually utilizing the iron for growth and genuinely multiplying? "One of our technical messes at the time, the differences between us all," Anna said, "was the problem of how we were counting mycobacteria. We counted in the cells, Mrs. Huang counted in the culture fluid, and the paper by the NIH authors counted them in the cells."

They did one of the Koch's-postulate experiments, however, reversing the procedure. They took the mycobacteria from some of the cultures that had for a week grown well in veal liver and injected them back into healthy mice to see if in fact they were viable. From August to November 1977, nothing happened to the mice at all. But, four months from the start, in December 1977, the mice were dying of leprosy. Using the liver had worked well; the mycobacteria certainly grew better than without it, and Anna felt that soon they could begin to concentrate on finding the correct medium.

Late in March 1978, eighteen months after starting his work, Nara, who had gone back to Rhode Island, turned up for a two-day visit, to pull some things together and to prepare the outlines of a paper for a big conference at Kiel where Anna was to present the work on leprosy. She was, however, not planning to focus on the question of culturing mycobacteria but on their initial question: Is there a genetic susceptibility to this disease?

Four people crammed into her office, a space seven feet square, partitioned from the laboratory and already occupied by two desks, one microscope, one table, and several bookshelves. To begin with, Anna recapitulated those observations that had been important to her ideas and would be to their future work: First, the marked difference in the degree of infection present in the lymph nodes in the body cavity and those in the groin; second, the three points at which she had noticed the mycobacteria grew best—the internal wall of the small venule at the point where it enters into the lymph node, within the macrophages that occupy the bone marrow, and in the T-lymphocyte areas of the spleen. From these observations she concluded once again that the environment of the site of infection is of real significance. "In future," she said, "I would like to try growing the mycobacterium in endothelial cells, taken from the internal wall of the venule. These contain iron in the form of an iron-binding protein, ferritin, and that might be the only reason why the cultures with liver extract worked so well."

As for the laboratory growing of the mycobacteria, she repeated that four people, Nara, Ann, Mrs. Huang, and Marie, had all separately managed to reproduce the peaks of the culture growth but were all facing the fact that no culture lasted longer than three weeks.

"So much," said Anna, "for that conclusion. The technique has defeated us amateurs. Enough for the moment is enough. It needs to be done really properly." There was, she went on, a possibility that an Iranian woman doctor, a specialist in leprosy, would be coming over to tackle the problem in collaboration with the group. Then she turned to the original problem: Is there or is there not a genetic susceptibility to leprosy?

A recently published paper by a certain Dr. Closs reported that his initial work with different strains of mice indicated that there *were* different genetic susceptibilities to leprosy infection. Dr. Closs had discovered to his satisfaction that one particular genetic cross produces a strain of mice more susceptible than the others. The leprosy arm of Anna's group had endeavored to repeat his work, and Anna's technician for the leprosy disagreed with Dr. Closs's findings, deducing an opposite result from their own data. Anna had then asked the biostatistics department for an accurate statistical analysis of their results, in order to see what valid interpretation could be put on the data. And they were now to discuss the matter.

"So," said Anna to the meeting, "we have done this very thoroughly, and, after four years' work, disgracefully we have come to a conclusion opposite to Dr. Closs's. We have absolutely the reverse conclusion."

"Why disgracefully?" I asked.

"I actually have no reaction at all," said Anna. "Results have to be explained. If other people didn't bother or worry about the fact that our results *are* different, I wouldn't be worried either. I have no guilt feelings. In any case, we've got in there. We've got our hands bloody at the bench. We are not *just* amateur technicians. We didn't just have somebody hand us the figures and say, Does it work? We were there sweating away, killing—how many mice was it, five hundred?" She was thoughtful for a moment. "Maybe we need a thousand."

The technician, Dee, intervened. "Over five hundred," she said, with great feeling, "and I quit."

Hastily Anna assured Dee that she didn't really mean it. She knew that the results to be gleaned from another five hundred mice would not in fact be worth the work involved.

But they still had to try to explain the difference between their results and Dr. Closs's. "Possibly we are talking about something quite different," said Anna. "But nevertheless, we've done solid work. So I will just say in a paper that our results *are* different. Did Dr. Closs count the mycobacteria?" she went on.

"No," replied Nara.

"There is one other difference between our techniques. He injected the infected cells subcutaneously. We injected intravenously. Maybe we are both right. Maybe what the differences in the results tell us—given that basically we did the same thing—is something about the mechanism of the infection." She picked up the data sheets again, and they all examined them in silence.

"Nara," said Anna, after a while, "there's a dreadful slump at one point—the death of all those B-6 male mice. Was it an accident in the cage?"

"No," replied Nara. "Absolutely not."

"Well, then, to a degree we do have identical results," said Anna.

"We used many more animals than Dr. Closs," said Nara. "A lot more. We used different sexes, and we had a different way of doing the bacterial counts."

"Well, then," said Anna, "we do agree with his structural findings. The mycobacteria turned up in the same places in both sets of experiments. O.K.? But I have another question: In what way—if any—do we agree with Closs beyond that?"

"Well," said Nara, "it is possible the different results reflect only the different ways in which we express them. We put down the difference in the survival rates and the bacterial index, and Dr. Closs put down the bacterial index only."

There was silence, and then Anna said, "Well, I'll tell you what this tells me. Listen. Listen. It tells me how important it is to look down a microscope—to look at *everything in each animal* separately. We did that in all five hundred mice. *We saw* the differences in the inguinal [groin] and the mesenteric [abdominal] lymph nodes, and in the bone marrow. And I tell you, this question about susceptibility and resistance to disease is a question not only between animals but between *organs in the same animal*. Closs may have variation in the genetic crosses, but we have variations in *one* mouse! This is more important than the gross genetics. And I'll go on. I believe that the principles which govern selective infection will turn out to be the principles that govern cancer metastases—I know that. The disease is worse in the lymph nodes in the groin; *this fact is significant*. It is telling us something. Can we treat these animals? Could we cure them with lymph-node cells from the abdomen, where the infection isn't so bad? Maybe in the abdomen there is something special in the environment that is resistant to the mycobacteria."

"You know," said Nara, "it has been known for a long time but has never been really said until now: In the very early experiments on living animals, scientists recognized that viruses had a proclivity for certain sites. But that was viruses; they didn't extend the notion to bacteria. But why ever not? That's why I liked your idea then, Anna, and that is why I like it now. You get a bacterium that causes meningitis in the lining of the nerves. But why don't we get a toe infected with meningitis at the same time? Then there is TB in the lung. The first time the lesion is always small. Then it goes to the lymph nodes in the chest. Why only there? They show," he said, with an air of great gravity, "these bacteria show 'lymphotropism.'" He coined the word with grave delight.

"I like the word," said Anna, satisfied. "That's what they do. O.K., Nara. Somehow we'll squeeze all this in at the Kiel meeting. We'll get it all in somehow. Nara, you may not know it, but you are going to be senior author on this paper."

This time the anticipated protest didn't come, and later the reason became apparent. There had been many changes in Nara's hospital since he had returned to Rhode Island. The pathology department had "moved away," said Nara, "from the old style of cleaning the meat to the new style, in which you earn your stripes by publishing."

That was why he was at least ready to publish and, if not to take on the mantle of Jenner, at least to assume the well-deserved mantle of senior author on a scientific paper.

"You see," he said, "I want to buy a new house and I'm going to need a mortgage, and for the mortgage I'm going to need a raise, and for the raise I'm going to need a promotion, and for the promotion I've got to impress my boss, and my boss is now a new boss who believes in publication; so I am therefore going to have to increase my publication list, and therefore to do this I'm going to have to write this paper. Then when it is all done I can live with my wife in my little home and go on doing pathology.

"And by the way," Nara went on, "did you know that there is a change in the direction of the effort at the NIH? They are now going to concentrate on diabetes rather than cancer."

"Oh, yes," said Anna, heavily. "I know. Because we didn't solve cancer in two years, off we go now and develop another military operation and try something else. They really do not, I think, understand science. . . ."

But she was still restless, looking for the answer to the difference in degree of infection between the lymph nodes in the groin and in the abdomen.

"Maybe," said Anna, "it is because they have different T-lymphocyte populations. You see, the lymph node in the abdomen is draining the gut and therefore is providing a different environment for the bacteria. We could test this. We could take out the lymph nodes from under the tongue and have them grafted onto other animals, or the same animal—put them under the kidney, by the body cavity. Then we can inject the bacteria, and then we can kill the animals, remove the organs, and see if there is a difference."

"At the end of all this time?" I asked.

"At the end of four years of hard, hard work, we have a result. We have a different result from someone else. That's all."*

---

*By February 1980, a further eighteen months' work had confirmed Anna's hunch about the difference in techniques: When they injected the infection under the skin of the mouse, Anna's group could duplicate Dr. Closs's work precisely. When they injected into the vein, their own original results were duplicated.

"It all adds up, though."

"Oh, yes," said Anna tiredly. "In the end, of course, it all adds up, but people don't really understand the scientific process."

"Do you think you're in a cul-de-sac?"

"Oh, no," she said, "not at all. It's only a matter of time and commitment. Somewhere, sometime, some person will grow the things and they will use some of the facts we found."

PART II

Any old iron? Any old iron? Any, any,
                any old iron?
You'll look sweet, talk abaht a treat!
You'll look dapper from your napper to
                your feet.
Dressed in style, with a brand-new tile,
        and yer father's old gray tie on.
Oh, I wouldn't give yer tuppence for yer
                    gold watch chain.
Old iron! Old iron!

Traditional song of scrap-iron
collectors in London's East End

INTERLUDE

Anna had returned to New York in April 1976. I returned in September of that year, and the first thing she did was to sum up for me her intellectual state as she perceived it.

"As I've said, by the end of the early sixties, we knew that lymphocytes were found in the blood and to an extent we also knew their sources. But there is one important element in this history: In America, people are trained as good scientists as well as good medical doctors. So that thanks to the influence of a single group—Bob Good's group—you immediately saw the linkage between the basic functions of the lymphocytes and disease. In America they grasped the relevance of lymphocyte function very early on! The business of the T-lymphocytes' being related to cell-mediated immunity—immunity that requires cells in order to function—and of the B-lymphocytes' being linked to the production of antibodies, all *that* was being immediately transferred and, literally, transplanted into our understanding of disease. This connection, I repeat, was made immediately in the States.

"By contrast, and with the exception of some groups in Switzerland and Holland, people who discovered things on the other side of the Atlantic did so in mice and rats. They did not make the direct connection with disease. [Four years later, this has changed.] The result is that the fundamental discovery and physiology of the circulation of the lymphocytes, Jim Gowans's original discovery, has in Europe remained on the whole confined to rats or to mice. People have not put it into any wide clinical perspective whatsoever. I want to do this, and this explains why I am in this country—which I think is quite an important point.

"Now the problems I am interested in are so terribly simple that I get embarrassed when I talk to people who don't know anything about the subject. We have these lymphocytes circulating all the time, and there are diseases in which we have considerable reductions in the numbers of lymphocytes in the blood: mycosis fungoides,° leprosy, and Hodgkin's disease for a start. On the whole, people have made the connection with function, i.e., if we have too few lymphocytes, we have a poor immune response.

"Now my fundamental question is: If a lymphocyte is not here, why is it not here? Is it elsewhere? There are these malignant diseases in which this question appears to be significant, and this explains my interest in Hodgkin's disease. Then, of course, the relevance to cancer is the next question. Scientists have now found that people in danger of having metastases—when the cells of a tumor break away and spread throughout the body—have low lymphocyte counts. Again, on the whole, they have interpreted this fact in one way: There are just too few lymphocytes. But I am interpreting it another way: The lymphocytes are not in the bloodstream because they are probably in the tumor or have been utilized by the tumor. This is really my fundamental problem.

"The relevance I find that this may have in Hodgkin's disease is, first, that hidden lymphocytes may be acting as pointers to the problem. So we probably could be diagnosing the cancer much earlier than we do at the moment.† The second point is that you get this phenomenon of maldistribution; therefore, you will get some cell populations of lymphocytes in the wrong place, and consequently the patients are more susceptible to a viral infection, or parts of the body are more susceptible. Perhaps we could dissect out these lymphocyte populations and see if we could correlate differences in cell populations with differences in viral susceptibility. Perhaps we could then pick up people who are in danger of having a certain kind of disease much earlier than we can now. So my concern is not only with the understanding of the disease but with the question of prognosis and early diagnosis.

"Let's take the question of prognosis. If we know that a lymphocyte behaves abnormally, can we—by monitoring its function and monitoring its return into the circulation—anticipate how the disease is going to progress? Is it the same with Hodgkin's disease and mycosis fungoides and leprosy? Here is a lymphocyte that goes to a particular environment and remains there, which it isn't supposed to do. What are the circumstances that cause the lymphocyte to go

---

°A skin cancer. There is no common name for this disease.
†Recent work in man has confirmed that blood lymphocytes migrate rapidly to tumor sites.

there in the first place, what are the results of its misplacement, why can't it get away, and what is the clinical response we get as a result? This is where I'm at.

"And there are links here, I'm sure, to other cancers. So, finally, I want you to meet a young man who I think is doing the most important work on leukemia in this place."

"What is it?" I asked.

"Well, he has found a factor in the blood of leukemic patients which seems to control the production of certain of the white blood cells. Since this cancer is due to an uncontrollable production of white blood cells, I think his work is absolutely fundamental."

We went into the laboratory next door, and there I met a chunky, quiet, intense young man whom I reckoned to be about twenty-seven years old. Anna introduced us, in virtually the same words with which she had prepared me. As she spoke, Edward looked both pleased and deprecating at the same time. There was obviously a great amiability between them. We shook hands and chatted for a minute or two. I didn't see him again for eighteen months.

CHAPTER 8

SURPRISES

Many scientific theories have,
for very long periods of time,
stood the test of experience until
they had to be discarded owing to
Man's decision not merely to make
other experiments but to have different
experiences.

ERIC H. HELLER

Bureaucratically, Anna's section may have come into existence in April 1976, but effectively any new unit cannot properly get under way until permanent staff is established on a permanent basis in permanent premises with fully functioning equipment. Though the laboratory was not in satisfactory order until November 1976, at least they were in their new quarters by September. Over the summer Anna and her senior technician, Carolyn, organized both the transfer into a new building and the ordering and installation of equipment. They moved into totally bare space—without even desks or chairs—and divided it so that everyone could have at least a separate corner. The experience guaranteed that the two women got to know each other very well.

Though very different in many ways, not least in their countries of origin, both women seem very English in their relaxed, uncomplicated attitudes. Born in London, Carolyn had been in America for nearly three years when she joined the group and had already spent nine months at the institute. She had really wanted to be an M.D., but a combination of circumstances thwarted her, and the claims of science set her on another career. She enjoys it; she likes to wrestle with the problems, but she wrestles, too, with her natural warmth and concern for people, a concern that could pull her away from the bench to do something in more directly human terms. Gentle, down-to-earth, and practical, she has a deeply buried streak of vulnerability that surfaces from time to time. In repose she is classically beautiful, with features of Victorian softness and charm.

The next person to arrive was Michael, who came to the laboratory from Glasgow. Tall, lanky, humorous, and worried, he was to grow a mustache within the year that had the incongruous effect of making him look younger and not, as he probably intended, older. His commitment was first and foremost to his wife and family, to whom he was devoted. As Anna predicted, Carolyn liked Michael from the first. He was a meticulous worker, and his worried caution provided a perfect counterweight to Anna's enthusiasm and that informed intuition which jumps ahead, bypassing intermediate steps which are explicit in her mind but, more often that not, unsaid. Michael is a step-by-step person who has to understand the logical progression in full before he is prepared to accept any theory or experimental technique. Equally he is a natural and superb expounder of ideas, able to explain them with economy and a wealth of apt analogy.

Over the next few years a variety of people were to join the staff, most as temporary birds of passage. (This is a standard feature of the current scientific scene, where transients drop in for a stay of a few weeks or months to work on a specific problem or technique.) Thus two Portuguese arrived, to return to Europe within eleven months. Nara was there too, of course, and at six o'clock every evening, after returning from the laboratories at Rye, would wander up to Anna's lab and rapidly release a stream of stories and jokes. The whole friendly group would chatter away like budgerigars.

With everyone shaking down so well, the lab rapidly developed its collective sense of humor, an indispensable factor. Soon the newcomers were spending much of their free time together, and by December, when the wheels of the lab were running smoothly and the work was pouring in, they were all there working away most evenings. But Friday evening generally found them and their families in each other's apartments, eating, drinking, and discussing science in an entirely irreverent fashion. This proved to be a healthy attitude in the year that followed, when, as Carolyn said, the exhaustion of the workload was greatly eased by Anna's approach to life, which is to insist that no one take himself or herself too seriously. The absence of a hierarchical structure and her insistence on open-mindedness helped too.

"I have worked in at least eight different labs," Carolyn said, "and in all but Anna's, technicians were just—technicians. Moreover, we had a preconceived idea of what results we expected, and we tended to discard the others. But she encourages us in open-mindedness. If our experiments work in accordance with our predictions, fine. But if they don't, that is just fact and should be considered and explained as such. For example, I was working with her at home one night when a phone call came in from a microbiologist who had been repeating Nara's work in order to verify his results. I said, 'Let's hope she has the right answer.' And Anna responded, very crisply indeed, 'It doesn't matter what answer she has. If it is so, then it is so.'"

Gradually the laboratory became a scientifically ordered chaos. The group needed to perform all types of laboratory work, and as the equipment accumulated and accumulated and accumulated they had to be very imaginative about fitting it in and arranging it sensibly. Parallel to getting things in order was the ordeal of adjusting to New York and the American way of life and, even harder, to the American way of science. Within days, Michael had made many friends, was a member of a rugby football club, and played on weekends, just as in Scotland. Adapting financially, however, was extremely hard. A salary that from the shores of Britain seemed huge no longer stretched far enough, and soon he and his wife, Margot, and their infant daughter, Juliette, moved to Roosevelt Island. Carolyn, too, when she first arrived, had to make the same physical and psychological adjustments, using an alarming portion of her salary to establish a reasonable quality of life.

But adapting to the American way of science was more crucial and more difficult for Anna than for any of the others. As head of the laboratory it was her responsibility to bring in those grants which, after the short initial period of support by the institute, would guarantee the continuation of the work, and more prosaically, the salaries of her colleagues. People's lives had been dislocated in order to join the staff; Anna herself had resigned a tenured position in Glasgow in exchange for an annual contract which might ultimately depend on her capacity to "bring in" money, even her own salary. She had rebelled against that uncertainty even before leaving Britain, when the full realization of the professional gamble she was taking began to dawn. "It is impossible to produce good scientific results in one year, and especially when you have to start up a new laboratory. Even if everything was already functioning, it is still far too little time for people to demonstrate what they can do. Five years is reasonable; three a minimum."

Not too many years before, the institute had guaranteed its new associates and its foreign scientists five years. In 1976 this was no longer true, and today the struggle for funds is so severe that, with very few exceptions, a sword of Damocles hangs over everyone: no grants, no jobs. So, as John Leonard has said, grantsmanship, as much as discovery, has become the art form of American science. The ensuing tensions can be severe.

"There are two populations," Anna said one day. She was not talking about lymphocytes. "You will find them in the elevators around the laboratory building." It had taken one year to put up the new building and, since construction was going on all the time, the place was swarming with all sorts of workers. "One population is big, physical, relaxed, happy, noisy, and laughing; the other is pale, serious, and tight-lipped. Its members ride in elevators in deathly silence, without a glimmer of humor on their faces, only expressions of total anxiety. This second population is the scientists."

She herself was to hover uncertainly and unhappily between the moods of both populations. By September 1976, she was already badly affected by her grant situation and nervous about her own "grant-getting" capacity in America. (In Britain she had never had any problem.) She was deeply troubled on her own behalf, but even more on behalf of those working with her. Yet, even so, one October day she took a deep breath and said, "I've decided to act as if money were no problem. Otherwise I'll go mad and I shan't do science."

"How much money do you have, for how long?" I asked.

"Well," she replied, "the finance officer told me the other day that if I dropped a technician or two, the rest of the group could go on for eighteen months. So I asked him, 'How long can we go on if we all stay?'

"He answered, 'Seven months.'

"So I said, 'That's how long we'll be here then. We'll either all go or all stay.'"

For the time being, she ignored the problem and picked up again the work on Hodgkin's disease. Within one month another small clue surfaced.

In collaboration with other colleagues at the institute, Anna had sent a paper to *Clinical and Experimental Immunology* on April 13, 1976. This presented the results of her work on the Hodgkin's disease patients whom she had studied while on sabbatical leave. She and her colleagues were generally satisfied with the preliminary results, which showed that the absence of lymphocytes in the bloodstream of these patients was due *not* to a quantitative diminution of these cells but to the fact that the lymphocytes were trapped, neither circulating nor migrating as they normally do. She was now ready to go to the next stage.

Thomas Hodgkin first described the disease which bears his name in 1832, citing seven cases, of which three would qualify under our present identification of symptoms. The first symptom is the same whether the patient is a child or an adult: a painless swelling on one side of the neck. Because children are generally watched with attention by their mothers, this swelling is usually noticed sooner in them than in adults. In any case, adults find all sorts of reasons for such bumps and tend to ignore them. Sometimes patients have additional symptoms: a fever which can be either fairly quiescent or really stormy, or night sweats, or loss of weight. But by and large half the children come in for examination with just a painless swelling. Usually these swellings are harmless lumps, but it is nevertheless important to check them because, as with so many cancers, the earlier the diagnosis the better the prospects. A worried mother is thus a godsend to clinicians.

Because overt symptoms are so similar to those of an upper respiratory infection, the final diagnosis of Hodgkin's disease depends upon a biopsy. There is very little else to go on. Immunological studies of the blood can be informative, and they do show that patients share some impairment of immune function. But even though one won't have Hodgkin's disease without having impaired immunity, one can have impaired immunity without having Hodgkin's disease. For accurate diagnosis the lymph node must be taken out and the final assessment made by pathologists, who examine its cells looking for one in particular: the Reed-Sternberg cell, named after the man and the woman who first described it. Its origin has so far eluded scientists. It is a cell that almost always has two nuclei, and sometimes more, but that is about all one can say. Where it comes from, what it is, how it got to be this way has remained a mystery. But since its presence is typical, its existence is *the* diagnostic feature of Hodgkin's disease.

The disease progresses through four stages, beginning with the mild enlargement of the lymph nodes in the neck. From the doctor's point of view, the stages are defined by the gradual encroachment of the disease into lymph nodes in different parts of the body; if the disease is confined to lymph nodes in one side of the neck, that is Stage I; if on both sides of the neck and above and below the diaphragm, Stage II; if the spleen too is involved, Stage III; when the disease is all over the body, that is Stage IV. The prospects for the patients differ according to the spread of the disease and so, too, does the treatment. In the early days—that is, the 1960s—clinicians were dependent on X-rays alone. So they injected a dye through a vein in the foot and then X-rayed the abdomen to see if the nodes there were enlarged. Now they do a laparotomy: They open up the abdomen, take out the spleen and those nodes that look suspicious, and biopsy them. With greater precision in the diagnosis, there can now be more precision in the treatment. The period before the 1960s was a sort of Dark Ages. Wherever the disease turned up, there radiation was concentrated. It was a cruel kind of Tom and Jerry cartoon: The disease popped up in one place and was slapped down, only to reappear in another, to be hit again.

The last few years have changed that procedure radically, even though, as in the case of many cancer treatments, clinicians walk a tightrope. If Hodgkin's disease is treated too aggressively, especially in adolescents, the treatment may harm the normal tissue, and if the patients live long enough they may develop a secondary malignancy as a result of the treatment of the first. Nevertheless, 90 percent of the children in Stage I now survive in a disease-free condition, which is quite remarkable progress. If caught early enough in Stage II, the figure is in the range of 80 to 90 percent. In fact, the treatment of Hodgkin's disease is one of the real success stories of modern cancer research.

The disease is widespread geographically; certain countries, and certain places in those countries, have a significantly higher incidence than the rest of the world. Beirut is one such place; São Paulo another. Where the disease occurs in Asia, it occurs most widely among the young. The United States has the largest number of cases, in both children and adults, but this statistic may only reflect better and more extensive diagnostic coverage.

The present world authority on Hodgkin's disease is, it is generally agreed, Dr. Henry S. Kaplan of Stanford University Medical Center. But before him Dr. Noel Craver was the figure of renown, and he, until his retirement, was attached to the cancer research institute where Anna and her group were working. There Dr. Tien-Chun, the clinician with whom Anna was to collaborate closely, was able to benefit from his presence and example. Ever since joining the staff in 1952, she has concentrated on Hodgkin's disease, inspired by Noel Craver's conviction that when we know this disease we shall know the whole of medicine because its ramifications extend almost to the entire medical field: to epidemiology, to immunology, to viruses, to malignancy.

Tien-Chun, now in mid middle age, left China in 1948 by what she describes as a series of miracles. In 1948 the Communists had not yet taken over, but many people were trying to leave. She had just $200—the boat fare to America—where she planned to continue her medical training. But a U.S. maritime strike was going on, and Tien-Chun had no money for the air fare.

One small miraculous incident after another finally brought in the money, and Christmas Day 1948 found her on the glittering streets of Los Angeles, crying her eyes out in loneliness and homesickness. With her last remaining $50 she took the train east, feeding on cheese and crackers for three days. Once arrived in New York, there she has remained.

Tien-Chun is deeply religious. Her husband, whom she met at the Chinese Christian Fellowship in New York, is a minister. The combination of a compassionate faith and her own triumph over adversity has endowed her with a serenity that she transmits to others. She is also very hard-working and has, I was to find, a wicked sense of humor. Although she can get angry, I have heard her say only one unkind thing about anyone, whether an incompetent Westerner or an uncritical Communist: "Dummies, dummies!"

The hospital has records of Hodgkin's disease patients going back fifty years—to 1929—and these constitute a gold mine of data and empirical information. Tien-Chun herself, who in the past quarter of a century has probably dealt with more children afflicted with the disease than any other clinician in the United States, has kept detailed and extensive records of every one of her patients. These records are started when, after a patient first comes into the hospital and a biopsy has confirmed the disease, the initial "work-up" is set in train. This is a complete physical and biochemical examination that takes about a week or ten days and yields some thirty pieces of biochemical and physiological information. Then the experts get together, the teams of surgeons, radiotherapists, and pathologists, to look at both the X-rays and the facts provided by the workup, so as to determine the stage of the disease. If the patient is scheduled for exploratory surgery, the same group reassembles after the surgery in order to reassess the situation in terms of past experience. All the protocols—the therapeutic procedures to be followed—are written well before treatment begins, and once a course of treatment is agreed on, it follows routinely.

In the case of a child suffering from Hodgkin's disease, the first three years after treatment begins are crucial, and from these years additional quantities of data are derived. Each time the patient is seen, further tests are given and more information is recorded. A child receiving radiotherapy will come in four days a week for three or four weeks and will be seen by Tien-Chun once each week, or until the disease has abated; then once every two weeks. If, after two or three months, the disease has not developed further, these checkups become less frequent. If, however, the disease has moved into a later stage, or if there is some physiological abnormality, the child must be checked more often and every change must be scrutinized and the new facts recorded.

Keeping records on such a scale is certainly necessary and interesting but vastly time-consuming. It can often take a whole day to do one chart, and Tien-Chun is actively involved with some 250 children. (Computer help with these records was to come much later.) Also, from 1971 onward she began to add extensive immunological data to the existing records. At the same time she reviewed the records of all her past patients. The system, once she set it up, continued automatically, and by now she believes that her files contain any kind of information anyone might ever want about these patients.

If one considers the thirty facts obtained from the first workup of 250 patients—a total of 7,500 facts—and adds to them the additional facts recorded each time one of the 250 has visited the hospital, it is clear that an incalculable mine of clinical information has been accumulated. Tien-Chun had no idea of what to do with this lode of data, a situation by no means uncommon for clinicians. From time to time colleagues would ask for access and it was granted, but she was unwilling to let the material out of her hands—for two reasons. First, no one ever gave her a clue as to what they hoped or expected to do with the data, and second, she felt a small ripple of mistrust, because she was never wholly convinced that the people asking for the material were really interested in Hodgkin's disease except in relation to their own careers. But in Anna's case she felt immediately things were different, because Anna came to her with a picture of what she was after and a theory that made some sense.

"Sure," Tien-Chun told me, "we were just jumping from here to there. But I was amazed that each time she came in with an idea, I was able to see exactly what she was thinking and where this might fit in the overall picture. So she was able to convince me very easily that she could do something with the data. Then I said, 'By all means, go right ahead.'"

To carve a way through a forest of empirical data, a scientist's tools are mostly mathematical, and of these tools graphs are the most important. From the thousands of pieces of information available to Anna, and to Dr. Ruth, another Chinese doctor who worked in a Long Island hospital, the most important were the immunological facts. It was on these that they concentrated, as they set about making graphs, plotting almost everything against everything else: blood counts; the different lymphocyte populations as measured in the blood; the amount of iron in the serum; the amount of ferritin in the blood; the dividing capacity—mitogen responses—of the lymphocytes removed from biopsied or treated patients (there were now three different variations of this test), all plotted against specific times; the week-by-week progress of the patients, including the dates when they were operated on and the dates of their treatments. It was not the kind of job one could sit down and complete in a week: first, because the quantity of data was so great that boredom and exhaustion would eventually result in loss of accuracy, or interest, or sheer stamina; and second, because Dr. Ruth, another of those meticulously exact people who loves playing around with graphs and figures, was not always free to come in from Long Island.

During the days and nights when they were working on the graphs, the two scientists would be submerged for hours, like a couple of deep-sea divers meticulously searching for treasure. Theirs was a slow, painstaking analysis. In the same way that underwater archaeologists may not skip from here to there but must survey the ocean bed inch by inch and area by area, they had to turn over the "stones" with great care, knowing precisely what stone they were moving and where it fitted. Like deep-sea divers, too, they had to come up for air, and if one strayed into the office late at night one would find them chewing mushoo pork or drinking bad coffee from the canteen. The piles of paper grew higher as the mountain of data shrank, transformed into curves and comparative columns and dots and lines. "Chopping trees, just chopping trees," was how Anna described it.

Soon, and sadly, they knew that nothing was going to stand out instantly and clearly. The essential clue to Hodgkin's disease and immunology—if there was one—was not going to be presented to them on a plate. Some evenings, they would shift their locale, and the dining-room table in Anna's apartment would be covered with pencils and rulers, graph paper and transparent paper, black, white and red letters of sticky letter sets, adhesive numbers, stippling, and shadings—all the modern appurtenances of scientific illustration. (These make illustrating a scientific paper far easier than in the days when you slowly and painstakingly shaded in your own graphs, just as you washed up your own glassware.) A job of such formidable dimensions has to be conceived in deep interest and critical need and executed with high enthusiasm—certainly with the conviction that somewhere in the mound of data a nugget or two is buried. The initial optimism may be great, but very soon the task becomes just one long, hard slog, and as the work with Dr. Ruth went on, there were times when the first morning light was on the East River as Anna walked home from the lab. But unlike her walks homeward from the laboratory in Glasgow in the hour of the dawn chorus, there was no song of blackbirds as counterpoint to the sound of her footsteps.

The climax came in mid-October 1976. One morning Anna joined Dr. Ruth at 8:30 A.M. and worked through until 11:00 the following night, with only a couple of hours' sleep snatched on a chair in the laboratory. And the very next day she said to me, "It is all pointing to something very simple."

The graphs were revealing many things, but one thing in particular had caught her eye and excited her imagination. Plotting the various immunological factors against the progress of the disease in each patient could not of itself reveal enough, so they laid the graphs for the patients one above the other to see if there were any patterns in these metabolic changes. Did the typical physiological changes all appear at the same stage in the disease, or did they precede or succeed one another? And they found one consistent pattern: The lymphocyte count in the blood went up *just before* the level of serum iron in the blood went down. This was mysterious and intriguing.

At this time a scientist from Israel was visiting the institute and gave a talk on Hodgkin's disease, emphasizing the presence of the iron compound, ferritin, on the surface of the lymphocytes in the blood of Hodgkin's disease patients. The changing amount of ferritin in the blood during a certain period of time was one of the facts plotted on their graphs, and at this moment Anna began to wonder if there might perhaps be a direct connection between lymphocytes and iron, or if the lymphocytes might play a critical role in some relationship between the iron cycles in the body and the immune system. She knew absolutely nothing about the biological effects of iron and only a little about the iron cycles in the body. But the regular and consistent pattern of the lymphocyte count going up immediately preceding a drop in the iron in the blood provided a target for investigation.

It has been known for a long time that the level of serum iron *is* low in Hodgkin's disease patients and that a direct correlation exists between this level and the extent of the disease: The worse the disease, the lower the iron content of the blood serum. (As the disease is treated and the patient progresses the serum iron rises again.) At the same time the crucial "mitogen response," the dividing activity of the lymphocytes, is also low. So there were now four facts to encompass: the ferritin on the surface of the lymphocytes, their lowered capacity for cell division, the low serum iron in the blood, and the increase of the number of lymphocytes in the blood just before the serum iron went down. Were the ferritin-coated lymphocytes taking up iron? If they were, should they be? Was this a natural thing to do, or were they doing it because of the disease? Perhaps the iron metabolism was all wrong.

Anna kept saying, "I'm sure there is *one simple thing* wrong. It may be a defect in metabolism—an error of some kind—a single enzyme defect of some kind." When I asked how she planned to go about finding out if this was so, and which enzyme might be involved, she said quite simply, "I shall just have to look at the whole field of iron metabolism, for a start."

She approached her newest hunch with a mixture of excitement, cynicism, ruefulness, and wry amusement. For the history of the struggle to understand Hodgkin's disease has been a record of one red herring after another, and she knew her hunch could easily turn up one more. But because no one really knew where to turn next, there was certainly nothing to lose. Nevertheless, she was rueful because she knew that if she persisted with this hunch, she would be challenging prevailing theory in two ways. The established theory— dogma, even—of Hodgkin's disease is that whatever the cause—and it is around this point that the red herrings swim in vast shoals—one cell becomes malignant. By division of that one cell, a whole colony—"clones"—of cancer cells derives which invades the rest of the body. But Anna's idea of an error of metabolism as a basic cause would mean that Hodgkin's disease would be a metabolic problem first and a malignancy second and might not result from the divisions—cloning—of *one* aberrant cell alone, because many cells would be simultaneously affected.

When a suggestion is first broached in science, however tentatively, it can be, and often is, bolstered by little pieces of information which up to that point may well have seemed extraneous. These can now be picked up and cemented in place. It is like creating a new design from existing odd pieces of colored glass. Indeed, Anna once gave a lecture called "The Stained Glass Window Lecture," explaining that all scientists have these little pieces of colored glass, intriguing bits of information or facts which they don't quite know what to do with. They leave them lying around until, prompted by a new idea or a new piece of information, they mentally sift them and select the ones that may help the pattern. In situations of conceptual drought one actively prospects for such pieces. So it was in this case.

Could there be a defect in metabolism, one small metabolic problem that might be the trigger that precipitated everything else in Hodgkin's disease? The metabolic pathways in the body are vast in number and most complicated to unravel. To start to go through them randomly would not only be scientifically pointless, it would take an eternity. The question was: Did they have any clue at all from the obvious symptoms of Hodgkin's disease that might suggest one metabolic pathway above all others that would be worth starting down?

Of all the obvious symptoms, only one seemed distantly connected with metabolism. It provided a very remote clue indeed, but at least it was a start, and Anna from the outset insisted that it could be very important. Very soon after drinking alcohol, some Hodgkin's disease patients experience intense pain at the site of the disease. But one well-known feature of alcoholism is an inability to cope with iron. In addition, Dr. Tien-Chun had reported that some of her child patients had very high levels of uric acid, and again nobody knew why. Now, from a preliminary reading of papers on iron metabolism and from a textbook on hematology that Nara had brought into the lab, the group learned that an enzyme, xanthine oxidase, is involved in *both* uric acid metabolism *and* the iron cycles of the body. Here was a second indirect link to iron. So Anna spent the whole of the next weekend in the library, tracking down and reading all the papers she could dig out on alcohol metabolism, on xanthine oxidase, and on iron cycles. Then another connection revealed itself: In one extensive survey of alcohol metabolism she found a reference to a study done in rats which showed the direct effect of alcohol on yet another enzyme, ALA synthetase. The suggestive point was that this enzyme is indirectly involved in iron metabolism too and provides a link in the chain of hemoglobin production.

This, too, was only a small clue, but Anna insisted that one intriguing link was enough and they should try to exploit it. She set Michael searching the literature on ALA synthetase and iron cycles, and he came up with a superb review of everything that happens biochemically in iron metabolism and of how scientists now think it works. Then they began to draw large diagrams of the "hem" pathways and iron cycles in the body—that is, the paths whereby hemo-globin, the main constituent of our red blood corpuscles, is created, synthesized, broken down, and utilized. Hemoglobin gives the red blood cells their color; it transports the oxygen; it is an iron com-pound. The oxidation and subsequent reduction of hemoglobin are what allow us to breathe, and the rate of production of hemoglobin is limited by the enzyme ALA synthetase. One by-product of the whole cycle is a substance called *porphyrin,* and defects in the pro-duction of porphyrin can lead to *porphyria,* a disease which pro-duces symptoms of madness. This is what George III of England suffered from, and now we know the cause: One crucial enzyme was missing.

So Anna and her colleagues began to look closely at those iron cycles, and most particularly at porphyrin. Was the problem there? How does the porphyrin reveal itself? They found a technique al-ready available which would help them. Porphyrins have a natural fluorescence, and therefore cells from Hodgkin's disease patients could—provided the cells were alive—be examined under the flu-orescent microscope without any particular preparation. One had to move fast, for when the cells are out of the body their capacity to fluoresce brightly under light of ordinary wavelength fades. (Fro-zen sections, though, retain some fluorescent capacity.) A darkroom is also necessary, and so they further crowded their already over-crowded lab with a curtain rail and curtained off with black fabric a small enclosure against the wall farthest from the window.

One day Michael reported that G, a patient in Stage IV of Hodgkin's disease, was to have a bone-marrow biopsy. By then Anna was at the receiving end of most human material, so that if specimens were expected which might interest her she was asked whether or not she wanted them. She wanted this one very much in-deed. That night, a Thursday early in December 1976, she looked at G.'s bone-marrow cells, alive and unstained, and saw a bright red fluorescence throughout. She wanted to stay working with the ma-terial, but she had an abstract to write and a deadline to meet. So she set some of the bone marrow on one side, having first arranged to have its cells measured the following week for porphyrin levels.

Before leaving the laboratory, she telephoned Tien-Chun to re-port what she had seen. There were, she said, enough positive flu-orescing cells to make her feel confident about proceeding, but she wanted to see as much material from patients as she could get. And right then she remembered that early in September, about the time Michael arrived, they had been given sections from the spleen of S.M., a young patient in Stage I of Hodgkin's disease. These sections had been frozen. She got the frozen sections out of the fridge, and I looked at them with her. In the spleen were areas full, chock-full, of fluorescing red spots. So immediately she rang Tien-Chun again and asked whether she considered the spleen of the patient, S.M., to be an involved spleen—that is, affected with the disease. Tien-Chun said she did. Once more Anna looked at the sections carefully. There were, she remarked aloud, odd macrophages in the sections, lots and lots of them. And suddenly she exploded:

"They're almost Reed-Sternberg cells. Oh, my God, I think we've made it. I think we've got the clue for the diagnostic cell. I bet you the Reed-Sternberg cell is just one step farther on in a sequence of changes in the macrophages. What I'm seeing here is not quite a Reed-Sternberg cell, but it will be. It will have just one more additional change in the same line. So far as the disease is concerned, we might as well forget everything we've ever learned and start again."

She rang Tien-Chun yet again to try out this idea on her. Tien-Chun's comment was, "For the first time in my life my mind is completely wide open about Hodgkin's disease. I'm looking at things I've never paid attention to before."

A second aspect of that evening's events also stirred Anna's wonder. It was, she said later, a perfect example of what Sir Peter Medawar has called the "anticipation of nature."

"I was looking at those spleen sections of S.M. three weeks ago," she said, "and I missed those fluorescing cells. I missed them because I was not looking for them. Now I am, and I see them. *But they were there all the time!* And what has changed? Only my thoughts."

Anna next wanted to examine the patients' blood to see if the fluorescence could be picked up there. If it could, that would be a great advantage, since it is much easier to arrange for a pinprick than an operation. On Tien-Chun's next day in the clinic, Anna went along and several patients were tapped for blood samples; three with Hodgkin's disease, one with chronic leukemia, and some with other leukemias. The very first specimen Anna looked at be-

longed to P.V., an adolescent girl with active Hodgkin's disease, and
it was highly positive for the strange fluorescing cells.

It was an extraordinary moment when Anna saw this blood. She
was alone in the laboratory. The case was one of the most difficult
Tien-Chun had ever had, for the patient was responding neither to
chemotherapy nor to radiotherapy, and here she was, highly positive for the autofluorescent cells. As the specimen slide was moved
around, these autofluorescent cells could be seen in every field of
the microscope. Anna was very moved, because two weeks before
she had quite literally not known a thing about iron metabolism and
had therefore to start with the textbooks and asking around. She had
suspected "something" might be there, but she was not at all certain
what she was looking for. Then she had started to search. The fluorescent cells she saw now were as exciting to her as a new super-
nova to a cosmologist.

"Of course," she said warningly, "I may be wrong because so far
these are all qualitative estimates. It is just an impression of quantity—and now we must begin to quantify accurately. So all the material will be tested."

The time of specimen collection began: a Hodgkin's disease
spleen and a non-Hodgkin's disease spleen; a Hodgkin's disease
bone marrow and bone marrow from a different sort of patient.
They obtained a lymph node from P.V., the refractory patient, and
Anna asked for additional frozen sections of it to be left unstained.
"And I bet you," she said to me, "they will be absolutely full of the
red stuff." They were; red and orange cells, looking like odd macrophages, were dispersed throughout the matrix. Lymph nodes, bone
marrows, peripheral blood, spleens of patients and of controls—
Anna looked at them all for the fluorescing cells and arranged to
have them all measured for porphyrin levels. So busy was the group
now and so restless, like hounds straining at their leashes, that their
formal and informal meetings, the grant applications to be prepared, the papers and reports to be written, all seemed ridiculous,
unnecessary and irritating intrusions into the exciting and vital
hunt.

One day, when they were grappling with the problem of the enzyme ALA synthetase, Michael joked, "This is Hodgkin's disease
project Number Three Sixty-five B. This time it really is ferritin, or
maybe ALA synthetase. No, that's project Number Three Sixty-five

A. That's how discoveries are made around here. We think things out and then we go and discover them."

"Dear Michael," said Anna, who was only half listening. "No wonder there are so many Hodgkin's disease projects! No one really knows *anything* yet. . . . It is a strange chemical, that enzyme," she mused. "That would be the beauty of it. It could be a single enzyme defect which altered the iron and the ferritin levels, and this somehow links up with lymphocytes and macrophages. Perhaps the macrophages are really knocked out with iron or something."

"But it has to be a malignant process eventually," insisted Michael. "The macrophages are so busy handling the garbage of the body that they must get really screwed up. The macrophages probably grab all the ferritin. They are not doing nothing all this time," he went on, in a bold flourish of double negatives.

"They are indeed," said Anna. "They are packing up the lymph nodes, stuffing them up. Tien-Chun always palpates the patients and only biopsies those lymph nodes that feel really hard and tough."

"But how does the porphyrin come into the story?" I asked.

"Well, the enzyme ALA synthetase affects the hemoglobin cycle, and porphyrin is a basic building block of hemoglobin. So it is involved, indirectly, in that way. If it were this enzyme that was wrong, it would be such a neat explanation—so neat that it probably isn't."

But, of course, they really didn't know. They were mulling it over and over, getting information from here, there, and everywhere, guessing about this and that, turning the picture in all possible ways. At times they were floundering.

Yet from various sources, evidence for a possible iron connection gradually built up—nothing conclusive, only suggestive. They found that many things affect the iron cycles. Phenobarbital is one; two of Tien-Chun's patients were on phenobarbital, and they came up with enormous lymph nodes. In porphyria patients, too, phenobarbital induces the production of ALA synthetase and the patients get much worse. Down the line in the iron cycle, yet another enzyme was important, one that increases in cases of lead poisoning. In fact, one Hodgkin's disease patient was for a long time suspected of having lead poisoning because of the amount of this enzyme is his blood. There was also a sex difference: Before puberty the incidence of Hodgkin's disease in girls and boys is very much the same, but after puberty the incidence rises dramatically in boys. Since in the blood of girls iron levels are reduced with the onset of menstrual bleeding, some sense could perhaps now be made of the sex difference.

Once one suspected a disturbance in the iron metabolism cycles, totally disparate facts began to fit. A perturbation at any stage in the cycle would generate a side effect that could actually mislead diagnosing clinicians. Other small clues either trickled in or could be set in place. The hunters began to look at the range of Hodgkin's disease throughout the United States, to see if any match could be made between the incidence of the disease and the iron content of the water or the proximity of iron industries. They matched up a map showing iron deposits with the maps of cancer incidence, and there did appear to be a possible correlation between iron ore and water reservoirs. (It was only suggestive and, in the absence of much more data, would remain just that.) Some fifteen years earlier, sixteen cases of Hodgkin's disease had been reported in one school in Albany, New York, but retrospective studies after such a period of time would prove nothing. Yet within the next twelve months similar outbreaks both in England and America gave them a chance to look again at these epidemiological links.

From her first observation that the lymphocytes went up as the serum iron went down, and in her desire to explain this, Anna had felt that she should look either for an abnormality of the lymphocyte which caused it to pick up the iron or to an abnormality in the iron cycles. "I don't quite know how this is going to link up," she said, "but it is good enough to put us on the hunting." I asked about the subsequent implications for therapy. *If* her ideas turned out to be true, she agreed, a much gentler therapy would be called for, intervening very sweetly at the point of the metabolic defect, rather than slamming away with radiation and chemotherapy. But it was—as always—too early to be talking about therapy! I was to come back in ten years.

Her first priority was to get precise values for porphyrin and fluorescing cells in the spleen and lymph nodes. "I am absolutely sure," she said, "that this is where the greatest amount of the stuff will be found. I really think ultimately you are going to have an abnormal macrophage, and the abnormality will be somehow connected with iron. I'm not too certain how this will work out, but this defect is either going to be expressed in the macrophage or in the red cells, which in turn would lead to the macrophage's having to eat them up, and more of them, and more of them. I don't know. But this fits in, you see, with what I thought before. The lymphocytes circulate, and one of the points of their interaction is with the macrophage. If in the lymph node you have abnormal macrophages laden with iron, the lymphocytes going there will interact with those cells. So the lymphocytes will be delayed and then will accumulate there. Then you will get a depletion of lymphocytes in the blood. And so Hodgkin's disease could have *nothing* to do with lymphocytes directly."

"That's a change from what you said last year."

"I know, I know. The maldistribution of lymphocytes could be just a reflection—just acting as a very sensitive probe to tell us something else is wrong. So where you find trapped lymphocytes you should be looking for *other* things that are wrong. For if lymphocytes are accumulating, there *must* be something wrong."

On December 16th, Anna's parents arrived in New York for a six-week stay, and that day the first figures came in on the amounts of porphyrin in diseased organs and in the controls: two bone marrows and two spleens. The amount of porphyrin in the spleen of the child with Hodgkin's disease was much higher than in the non-Hodgkin's case; similarly, the amount of porphyrin in the bone marrow of the man with Hodgkin's disease was much higher than in the other. The results were telephoned to Anna just as she was leaving for the airport. She grabbed a piece of pink paper, wrote the results down, and handed them to Michael for safekeeping.

Those results seemed of sufficient interest to warrant another meeting with Tien-Chun and other colleagues. It took place on January 6, 1977, and from the start it was clear that the meeting was going to be a tough one. The director was in a thoroughly skeptical mood. There were no trumpets at all. He said at the outset, "I might as well tell you: I firmly believe that Hodgkin's disease results from a defect in one cell which then clones." That flat assertion established the tone of the meeting.

Anna came straight back: "Well, O.K. This new idea may turn out to be another red herring, but at least it is the only one that is really red."

Anna, Tien-Chun, and Michael now reviewed the logic of their thought and the experience of the findings, both experimental and from the literature, reminding the director of a well-known fact about iron: Once in, it is extremely difficult to get out. No matter how much keeps coming, the body keeps absorbing it. We need some iron, but not too much, and if we have too much it must be dealt with somehow—otherwise, Anna asked, why don't we all turn into nails? They reviewed not only the data but the process whereby they had decided to look at the iron metabolism cycles. When they started to discuss the critical enzyme, ALA synthetase, and its relation to the metabolism of porphyrin, the director said, "Aha."

"But," Anna said, "wait a minute," and went on to emphasize that the weight of the accumulating circumstantial evidence seemed to justify continuing along these lines. Everything they had heard, seen, or read had begun to point toward a metabolic problem which somehow involved iron. She emphasized, however, that at this early stage they had no tight controls.

The director picked up his bucket of cold water again. "I don't believe a word of it. I'm still one of those men who believe in Hodgkin's disease as pure cancer."

"All right," retorted Anna. "Then the title of my presentation is 'The Ultimate Red Herring.' And yet . . . " And off they all went again, back and forth between evidence and objection, logic and suggestion, intuition and fact. And though the meeting took on the quality of scientific slapstick, the director was not totally dismissive. The next step was to sit down with a real porphyrin expert, and there weren't too many of them.

"Some people have done some work on these problems," said Anna, "but there are no funds for them."

"Oh, yes, there are," he retorted. "For the good ones. But the real problem, as I see it, is this: If you really have cancer in this Hodgkin's system—if you do have a malignancy—then all that you are seeing may be simply a perturbation of the system. Everything you are finding and looking at may happen as a consequence of the cancer. It is just as likely that instead of having an inherent abnormality of iron metabolism as the *cause* of Hodgkin's disease, an upset iron metabolism is its *consequence*. And as for making predictions, there is a Catch 22. If I'm right, cured patients would not have any abnormality. But if patients do have your basic abnormality they couldn't be cured."

"But we can still make a prediction," argued Anna. "Certainly, if your interpretation is right, cured patients should not show any abnormality; if I'm right, cured patients would still show an abnormality, but it could have been corrected or compensated for."

As they went back over all the material for the third and then the fourth time, the director recalled an old friend and colleague at Minnesota, Dr. Sam Schwartz, who had once wondered about the role of porphyrin in Hodgkin's disease but had obtained so little evidence from patients that he never developed the work further. Once more the director reviewed the clinical situation and once more came back to the basic objection: Hodgkin's disease occurred because of a combination of two things; a lack of cellular architecture and a sick cell line. Thus all Anna had found and other people were finding could be considered a consequence of these defects. In any case, he went on, arguing against the thrust of her thought, treating Hodgkin's disease patients by changing their diet doesn't work; there is no statistical association between Hodgkin's disease patients and thalassemia patients—people with a genetic defect that affects their hemoglobin synthesis. And if there was an iron connection in Hodgkin's disease, one could expect such an association.

"There will be one," Anna interjected.

At which point, musingly, the director returned to Dr. Schwartz. "Actually, he often thought there might be something in this iron and Hodgkin's disease business. There was a sense—a feel for some kind of relationship—picked up in Minnesota between 1947 and 1953. But he really never went ahead with it. What do you propose to do now?"

"Well, for a start, to flood the system—the lymphocytes in animals and other cells—with iron citrates and see what effects we get."

By the end of two hours the director had at least conceded this much: "You have a very interesting set of observations." And that was that.

Later, Anna said, "What cheek I've got! What a nerve! To suggest that Hodgkin's disease mightn't be a cancer first. But I don't really care *what* they think provided we can treat it. But when, oh, when, will I make my point? I'll be old and tired before I do. It is lonely being a scientist. Meantime we have a man who worked in an iron factory and he has an eight-pound spleen. Vast!"

From then on they were set to focus on the iron connection in three ways. First, it was essential to begin getting accurate measurements of porphyrins in the spleen and the blood; second, they would examine as many blood samples and spleens from Hodgkin's disease patients and non-Hodgkin's disease patients as possible, to see (a) if the characteristic fluorescing red cell was limited to the Hodgkin's patients, (b) in what quantities it was there, and (c) what kind of cell it was; third, they would all undertake thorough searches through the literature, to pull in as much knowledge as possible.

Immediately they ran into major problems. There was a genuine scientific difficulty. When Tien-Chun took one sample of a Hodgkin's disease spleen to be measured for porphyrin by a colleague in another institution, the measurement was made happily enough. But further measurements were refused on the grounds that the results were meaningless in the absence of normal human spleens with which to compare them. Scientifically, this was perfectly true and, scientifically, the only attitude to take. "But where, oh, where," said Tien-Chun in angry despair, "am I expected to get fifty normal human spleens?"

This is a constant and worrisome problem for people working on any disease that involves the spleen. The only consistent source of normal spleens is the emergency ward of a hospital, where one finds people who have been in accidents and whose ruptured spleens must be removed. But everyone wants a normal spleen; the material is at a premium and is very difficult to obtain. Though some nondiseased spleens did eventually get through to Anna's group, for the time being that particular route was closed.

Still, the porphyrins *could* easily be measured in the blood of patients and in controls. But here a second difficulty arose. Scientists are human beings with quirks, prejudices, loves, hates, passions, pride, and sensitive egos. After the first round of measurements, the scientists in another lab who had been doing the preliminary estimates suddenly refused to do any more, either because of hurt feelings or injured pride or emotional turmoil. Thus that work, too, temporarily ground to a halt. Anna was deeply distressed.

"We have so little time," she said, "that sometimes I long to be dead. Now you see what science is and can be. There are kids dying out there, and this kind of nonsense is waste, waste, waste. They won't do the tests and there is this six-year-old sick boy, his spleen and his liver.

"If necessary," she continued, "I'll fly out to Minnesota and do the tests myself. The kids are far more important than my feelings.

"Thank God," she went on, "for those people who keep a lab sane. I've never really been in those mad, take-off places, but I've very often been in places where there have been slightly insane people. Not insane intellectually, but insane in pettiness. Generally they are the very sensitive types. But having quiet technicians and completely bland people around is wonderful. They don't notice anything wrong. They just smile and they keep happy. They keep us all sane."

They now had to think of a way of doing the quantitative estimates themselves, although they were not set up to do accurate biochemical porphyrin measurements. Anna reverted to her old well-tried technique: counting with her eyes. Hour after hour, and day after day, she and Michael would count—scanning the microscope field, looking at all the cells in the spleen and blood samples, and measuring the frequency of the cells loaded with autofluorescent material. It was painstaking and time-consuming work, but by now there was a crazy intensity, a self-propelling momentum in the laboratory. They were all driving themselves hard. In between the usual routine of meetings, telephone calls, requisitions, articles, reading, and grant applications, every spare moment was taken up with counting, counting, counting. Once more, laboratory days flowed into laboratory nights, and throughout there was another difficulty to contend with, the worst difficulty of all. They were besieged by skepticism.

This is always to be expected, and in their case it came from within and without. Inside the group, two visiting scientists were totally opposed to the whole thrust, one of them having told Michael shortly after Christmas 1976 that it was foolish to spend any time on iron metabolism. Everyone, she said, knew that the spleen broke hemoglobin down to iron and protein, so the iron they were seeing was almost certainly just a by-product of normal processes. Similarly, within the larger environment of the institute, they had sensed a current of generalized opposition since the very beginning of the project—a faint hostility which they could feel during the various meetings and conferences. Time and again the group had clashed with the young physicians, who greeted with barely concealed derision any remark at all theoretical about the disease in question. And all Anna's grant proposals were faring disastrously.

Depression and confidence vary with well-being. They were all exhausted, yet they would neither stop nor let go. They were a group of fierce and cheerful, depressed and obsessed terriers. Every specimen that came in they took. They jumped at every opportunity to examine material that seemed important, no matter from what source or at what time of day it arrived. Many of the laboratories in the institute have a notice on the door which says, NO CLINICAL MATERIAL ACCEPTED AFTER 4:30 P.M. This group took everything whenever it came, and since they needed to examine the specimens live and at once, they did so no matter what the time. But the time always seemed to be four thirty on a Friday afternoon. Later in the summer Carolyn recorded one typical Friday, which provides a prime example of those hectic days between January and April 1977, a pattern that was to continue for nearly ten months.

9:00 A.M.: Our new colleague is already hard at work, continuing a previous day's experiment and setting up a new one.

10:00 A.M.: Michael and/or I stagger into the lab, full of coffee and Danish pastries. We take thirty minutes to adjust to the realities of the day, during which the following occurs—

10:01: The research nurse calls to inform us that a splenectomy is presently being performed and we are to expect the spleen any time.

10:05: Laura [a volunteer] pops into the lab, full of happy energy and refreshing information about the film screening/rock concert/ballet that she attended the previous evening.

10:07: Phone call from secretary. Please pick up reports.

10:08: Phone call for an insurance company. It takes five minutes to convince the caller he has the wrong number.

10:14: Phone call from Public Relations Department. Please expect film crew and television cameras.

10:15: Phone call from husband. Needs money to buy coffee.

10:16: Phone call from Anna with some instructions. She is spending the morning at home since she was up all night writing a paper for presentation at a meeting in Chicago (or Madeira, or Oxford).

10:30: Phone call from Pathology Department. Please pick up slice of spleen which is now ready.

10:31: Go over to Pathology Department and squabble fiercely with pathologists over size of specimen. I don't think it is big enough.

10:40: Return and start work on spleen, needling it apart to separate the cells. Send Laura on many errands—to the photocopying department, the secretarial services, etc, etc.

11:30: Laura finally concedes that she will have to help with the "revolting specimen."

12:00: Mei arrives—another helpful technician—and immediately starts to work, freezing, cutting, and staining specimens.

12:05: A colleague—a visitor—comes in for her share of spleen.

12:30: I'm hungry and, if I'm lucky, go out for a sandwich.

1:00: Return and work again on spleen. There is frantic activity going on all over the laboratory and no room to do it in.

2:00: Anna arrives to attend an important meeting.

2:15: Salesman tries to sell me photometer, scintillation counter, fluid counter, and himself.

2:30: Life is somewhat calmer, as the work gets done and we begin to relax a little.

2:31: Phone call from the research nurse. Some lymph node of the patient is now in pathology. Would we please pick it up? I send Laura off again.

3:00: More running around.

3:30: We calm down a little and start now to work on the lymph node.

4:00: We go and get some coffee.

4:15: Cry from Michael: "Here it comes: four fifteen on a Friday afternoon. You can set your watch by it." Al, the sweet messenger from Pathology, smiling happily and carrying yet another spleen specimen, enters the room serenely unaware that he is the most unpopular man in the institute at that particular time.

4:16: Throw hysterical fit!

4:17: Calmed down by exhortations from Anna to do some Yoga and then face up to it.

5:00: Settle down to work for the entire evening. We are all trying to monopolize the fluorescent microscope simultaneously. Many "Oohs" and "Ahs" of excitement as the specimens of the day are examined.

6:30: Fortified by coffee and sandwiches supplied by Anna. The work is going well. Michael is waiting for his share of the cell suspensions I have made; a visiting scientist produces her cell samples for the mitogenic stimulation studies.

9:30: Finished! Carried out of lab by husband and begin to think about dinner.

One Friday the entire routine was repeated for the nth time. But that was the evening when Carolyn had planned a special dinner party for the group. By the time they got to her house they had just one hour together before those with small babies had to depart to relieve baby-sitters. She was very upset. It was all too much. Michael too began to doubt whether one human spleen was worth such a disrupted life.

By March 5, 1977, they had accumulated results from fourteen spleens and ten lymph nodes from a wide spectrum of patients. At that point they were looking much more carefully and quantitatively at the tissues, and Anna was beginning to concentrate on the macrophages. I myself had been shown big blobs on several occasions—big "things" somehow deep inside the macrophages with which the macrophages couldn't cope. The stuff, whatever it was, was clogging the cells, whereas in a normal spleen everything is "eaten up" perfectly. Did she, I asked Anna, know very much about the macrophage and macrophage cycles? No she didn't; she would have to talk to Dean James Hirsch of the Rockefeller University, for it was in his lab that much of the earlier work had been done.

Anna had, of course, been following the work of Dr. Henry Kaplan at Stanford, studying his papers very carefully and reading between the lines. It was clear, she felt, that he too was thinking that the defect was in the macrophage. In fact, his laboratory was developing cultures of spleen macrophages, as Carolyn was doing, and he was beginning to focus really hard on that cell. This independent evidence that they might be onto something genuine was comforting. Certainly it made Michael happy, because he was now seeing those large macrophages with fluorescent bodies in the thymus glands of the diseased children.

"But," Anna emphasized, "nobody—nobody else would be seeing exactly what we are seeing now."

"Why not?" I asked.

"Well," said Anna, "we are seeing these things free and clear *because we are not doing anything to the cells.* We are just cutting sections and looking. Everyone else *fixes* the cells, and the fluorescence is lost."

Early in the spring, Anna was due at a conference in Chicago. In respect of that meeting, she was, she felt, in a situation like the one she had been in so many years back in Bern, when she gave her first formal paper about the thymus-dependent areas. The temptation was the same, "to let them really have it." She had worked on her new idea for nine months and, she said, "I think I know what the defect is, or at least I am near enough to know." But how should she handle the matter at a meeting of the American Association of Pathologists? Dare she go straight in and say that she was confident that Hodgkin's disease arose from a defect in the macrophage due somehow to its inability to handle iron? At Bern she had made a vivid impression with a similar unambiguous directness, combined with hard evidence. Some people had admired her for her firm conviction; others were furious. But no one had ever forgotten. She had been "almighty confident" then. Now she was confident but not "almighty confident." "When you measure things in test tubes," she said, "and you see all the numbers, you must always have doubts because so many things can go wrong. But when you have a fresh piece of tissue, fresh from the patient, and look at it down the microscope immediately, you know *nothing* can have gone wrong and you can be absolutely confident. So I am confident about these red fluorescing cells. But perhaps not absolutely confident, because maybe the bloody things really are in normal tissues too! That is the only point where I am not confident. I haven't yet seen enough normal tissues. We need normal human spleens and livers very badly. But I suppose we can do it in the mouse."

So they worked on. If the others had private anxieties about the work, I did not know about them. Anna, apart from her one reservation about controls, had none about the work, but they all had worries of other kinds. Carolyn was pushing herself far too hard in order to complete an M.A. thesis and the examinations which would set her toward a Ph.D. Michael was struggling with domestic concerns—a new baby on the way and the difficulty of maintaining a family in tolerable circumstances in New York. Anna, in spite of her resolution taken six months earlier, was very worried by her inability either to get grants or to write a grant application in the required format—one that demands total confidence in what you think you will find from your experiments—or, in fact, to be taken seriously at all. Her worry was whether in America her success as a scientist would be judged not by what she found and discovered but by her capacity to garner money. The financial situation combined with other people's skepticism really worried everyone, and the atmosphere became as variable as a barometer in uncertain weather. The science was exciting and eminently worthwhile, but the external pressures were shaking their equilibrium.

They were at the start of what came to be known as Anna's Blue Period—and not only because of the "blues." They had stepped up the pace and started a new routine of staining all specimens for iron deposits to see if the metal was in the cells. When stained, the deposits appeared deep blue.

One patient's lymph nodes were found to be chock-full of blue stuff. Carolyn was finding it in the blood samples too. "Now," said Anna, "it is going to be like chasing a butterfly—a blue butterfly. I think blue; I see blue; I want blue." She laughed. The lab was a persistent buzz of seething work. Nara was deep into consolidation of the leprosy cultures, and his well-known predilection for a quiet life was the focus of a good deal of teasing, to which he never failed to rise.

"Hurry up," Michael said to him one day. "Now hurry up. It is only another month before we can recommend you for the Nobel Prize."

"But I don't *want* a prize," Nara replied, agonized.

When Anna finally left for the Chicago meeting, she had decided to be a cautious John the Baptist. She would talk about the phenomenon she called *ecotaxopathy*, the diseased condition associated with lymphocytes in the wrong place. But she would talk about it only as a pointer to the truth. She would show how it works, and how following up the idea of the lymphocytes had led her to look for an abnormality in the spleen environments, and how *that* had pointed them toward looking at the autofluorescent "things" which they now found in the blood and spleens of the patients, and how all this had led her to suspect that something might be wrong within the iron cycles.

Just before she left she had been examining sections from another patient, a boy whose spleen had confirmed Reed-Sternberg cells. These were very important slides, and the whole group was very excited about them. Looking at the fluorescence in the boy's spleen sections, Anna said, "I'm paralyzed. I'm just paralyzed. For one thing, the virus is going to be the biggest red herring in the history of this cancer. And everyone in the house of science is just standing in the corner facing the wall. And here we are. We think we really have found out what it is. For years all the nudniks have been staring at the disease with fixed gaze. Here we've been really working for six weeks and we've seen something. And do you know why we've seen something?" she asked. "Because we've *thought*. I'm a bus," she went on, "not a tram. There is a certain logic in making a tramline discovery. But the real excitement comes when you jump the rails and make a new track. Not easy. The problem is to get rid of the constraints of your past. I begin to wonder whether scientific free will really exists."

CHAPTER 9

ANNA'S
BLUE PERIOD

The scientist who finds hypotheses must build over them
a grand edifice that can contain them.
JEAN PIAGET, The Mission of the Idea

Just about the time Anna got on a plane and flew to Chicago, Carolyn and Michael got into the institute's limousine and drove to Kennedy to meet a new colleague. Nozaki, a medical doctor from Ja-
pan, was thirty years old when he came to America with his wife
and two small children. He had first written to the director of the
institute asking if he could visit for a period of research, and his letter had been passed on to Anna. He came because he knew the level
of science was very high in the States and he would find good people with whom to work. He had another reason, too: "If I live in Ja-
pan forever until I die, I cannot have a great view of everything. I
cannot learn a different way to live or think about my life. I like to
experience different cultures."

Even as a medical student Nozaki had been interested in immunology, and after graduation he had chosen to work in a department
whose main concern was with autoimmune diseases—those in
which an individual's immune system turns on itself and becomes
destructive. He had worked on clinical immunology and therefore
had considerable experience in those techniques that measure the
functioning of lymphocytes in human beings. He knew how to isolate the lymphocytes from the blood, and he knew how to do the
classical test for detecting the different populations of lymphocytes.

This test is standard: If lymphocytes from a human being are
mixed, at a temperature of 4 degrees centigrade, with the red blood
corpuscles—*erythrocytes*—from the blood of a sheep, the sheep's
red blood corpuscles will stick to the T-lymphocytes like petals
around the center of a flower. This is a characteristic "marker tech-

nique" used to detect the presence of human T-lymphocytes. Immunologists call the phenomenon "E-rosette formation." Nozaki, more delicately, calls it "Making sunflowers."

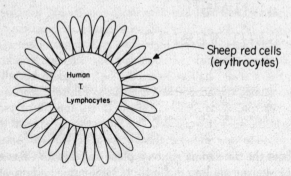

Making   Sunflowers

He found the phenomenon very strange. It seemed to work only with the sheep's red cells. Why on earth should this be so? What was the special connection between the human species and the sheep? He wondered whether the blood of other animal species would also yield "sunflowers" when mixed with human T-lymphocytes. He decided to live dangerously, to go a little wild, to leave the safe world of laboratory animals—guinea pigs, rabbits, rats, and mice—and try the test in a variety of animals. He found that it worked with dogs. He did it with seven or eight fairly mundane animals and wanted to try with others. He said, "So I consulted with my professor. He knows people who are in zoos in Okayama City. My professor is physician, and one of his patients is daughter of Japanese Emperor. Many years ago she suffered from sepsis. At that time my professor was her doctor and treated her. She has private zoo in Okayama. So when I consulted my professor, he said, 'O.K. I recommend you draw some blood from other animal: rhino, lion, tiger, elephant, or giraffe." [Nozaki pronounces this with a hard "g."] Professor said, 'I will say to daughter of Japanese Emperor to give you rhino, lion, tiger, elephant, or giraffe.' "

But at this Nozaki had second thoughts. " 'No, no,' I said. 'I don't know the ways of drawing blood from such animals. I cannot

draw blood from lion. It is dangerous. I cannot draw blood from giraffe. It is difficult.' "

And with that, I realized, we were back once more to the mouse,
the rat, and the old guinea pig.

Carolyn and Michael had firm ideas about what their new colleague would be like for they had seen numerous Japanese workers
around the institute: five feet high, thin, and almost noiseless, working from dawn to dusk without stopping. When Nozaki came
through the arrivals' hall at Kennedy, he proved to be tall, well-
built, handsome, and accompanied by an exquisitely beautiful wife
and two very pretty daughters, Akikoh, three, and Rieko, two. The
welcome at the airport went well, but when the car crossed the Tri-
borough Bridge and entered the outskirts of Harlem, small talk
about the flight faltered. Nobody said anything, but Nozaki and his
wife were gazing in horror at the burned-out buildings and the
trash-lined streets.

The institute maintains an apartment building for its scientists
and visitors, but "maintains" is a misnomer. Michael had moved his
wife and family out of the building very fast. When the Japanese
newcomers arrived, the apartment was filthy, the furniture shabby,
and there was only one bed. To make matters worse, the heavily
barred windows looked directly onto a derelict black building
which cut off all the light. The two little girls took one look and
burst into tears. Luckily a compatriot lived in the building. She was
appealed to and immediately took charge. Soon everyone felt a little
better, but as Nozaki later wrote me:

> The town was terribly more dirty than we expected. We must
> notice much faeces of dogs on the ground. It is incredible things be-
> cause there are few faeces on it in Japan. We had to take a walk look-
> ing downwards very carefully. The apartment was one bedroom. It
> seemed like one prison for us all.

They didn't last there long. With enormous energy, Michael organized their move to Roosevelt Island where he himself was living and
where Nozaki and his family are still.

Two days after they arrived, Anna returned from Chicago, but
there was time only for the briefest of welcomes before she was
scheduled to go off again, this time to Poland via Basle, to join Se-

bastian as guests of the Polish Academy of Sciences. Nevertheless early in April, just before she left, the group met to assess just where they were and discuss the problem of the iron cycles all over again.

It was then I learned that there are three major iron-binding proteins: *ferritin*, which we have met; *transferrin*, whose function is obvious from its name—to transport iron from one point to another; and *lactoferrin*, an iron-binding protein which we get first in our mother's milk and which soaks up iron like a sponge. In the midst of the discussion, Carolyn said, staring at the diagram of the iron transfer cycles, "Maybe we are not paying enough attention to transferrin."

"You may be right," Anna said.

Eight weeks later, when Carolyn came back from a holiday, "Everybody," she said, "was yelling about transferrin."

About one hour out on the April 20, 1977, British Airways daytime flight from Kennedy to Heathrow, Anna asked one of the stewardesses for some writing paper. I was on my way to Europe on the same flight, and fifteen minutes later she showed me the letter she had written:

Dear Nozaki,

I am sorry to have left you without a clear-cut subject to work on.

As the miles go by over the Atlantic, the more I think about T-cells having a receptor for Fe^{+++} [ferric iron] on their surface, and also ways of testing the hypothesis.

Well, it has just occurred to me that you are in the best position to test it; the sheep erythrocyte might have something unusual about its surface Fe [iron] to explain the reason for the spontaneous rosetting.

The experiment to test is shown on my sketch. [See art on opposite page.]

Well, if this is correct, if there are, as I think, receptors for either transferrin or ferritin on the human T-lymphocytes, then the antisera should stop the rosette formation. Moreover, conditions that change Fe^{+++} to Fe^{++} [ferric iron to ferrous iron] should also reduce rosette formation, i.e., pH, calcium, etc. Would you like to do some very simple experiments with sheep erythrocyte rosettes? We have the antisera (ask Michael), and you can modify the pH very easily. In theory an alkaline pH would decrease rosette formation, and an acid increase it—

A. If there is a receptor for ferritin on the surface of
    the lymphocyte

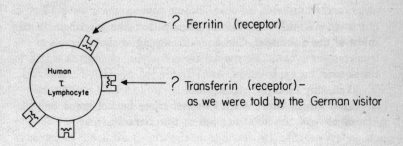

Then

B.

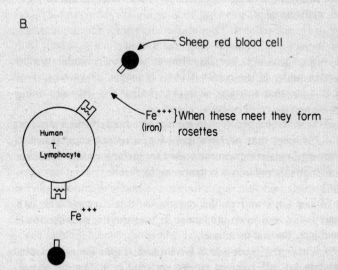

also doing the test in different oxidation conditions, etc. . . .

I'll leave it to you. Carolyn will help you to order the sheep erthro-cytes, or you can borrow some to start with from another lab.

Good luck!

Love,
Anna

P.S. An even better experiment would be to take some labeled Fe59 [ferric] and measure the uptake by T-cells in the gamma counter.

I needed an explanation. Anna reminded me that Nozaki had arrived at a time when the group had begun to set up routine stain-ing procedures on all materials—whether sections of tissue or cell suspensions—for the identification of iron deposits, whether ferric iron or ferrous iron. (In ferric iron, the iron is oxidized; as ferrous iron, it is reduced—that is, converted to another form.) By then Anna was seeing blue deposits everywhere, around all kinds of cells, but try as we might, neither Carolyn nor Michael nor Nozaki nor I could genuinely confirm these sightings. But two days before leav-ing, when they were all growing thoroughly skeptical, some blood had come in from a patient. They stained it in their routine manner, and when Anna gave her usual yelp and started shouting about blue spots, everyone groaned. Yet this time it was unarguable. We all could see that many of the cells had clear blue iron deposits on their surfaces. But whether this was normal and what the iron was doing there were questions still to be worked out.

It had been the surface appearance of those blood cells that made Anna suspect that perhaps there was a receptor for transfer-rin, the iron-transporting protein, on the surface of the lympho-cytes, and that this protein was trapping the iron.

Having explained this much to me, she reached into the pocket in front of our seats and pulled out the British Airways sick bag. Above and below the instructions for its use, she drew diagrams to illustrate what she had in mind.

She drew the iron cycles: Iron is oxidized; it goes round as trans-ferrin; it is stored as ferritin; it is then reduced and is bound to por-phyrin, a building block of hemoglobin; then it becomes part of he-moglobin. The role of transferrin is to carry the iron in one form to places like red blood cells, where it can be stored as iron in another form. And now Anna was certain that the T-lymphocytes would prove to have a receptor for iron, or iron-binding proteins, and were somehow involved in the cycle.

Thus the letter to Nozaki meant: "Making sunflowers" with sheep red blood cells is a test to reveal the presence of T-lymphocytes. If, as Anna guessed, transferrin is the substance which causes the T-lymphocytes to cling to the sheep's red cells, then, in the presence of additional iron, this process would be stopped because the transferrin would pick up the iron instead, latching onto the metal rather than the sheep cells. The process could also be stopped by using antitransferrin—a substance that smothers transferrin just as effectively as antimatter smothers matter. Thus the series of experiments she was suggesting would give a clue to what was on the surface of the lymphocytes and indicate if they had a receptor for iron.

However, there was more to her letter than that. It was very valuable to have someone around—Nozaki—who had a scientific interest in understanding the phenomenon of "making sunflowers"—the E-rosette formations—with T-cells. Several groups of workers had already found that the number of T-lymphocytes that react to "make sunflowers" in the blood of Hodgkin's disease patients is reduced by approximately 50 percent. Perhaps there really was "something" in the serum of Hodgkin's disease patients that attached to the lymphocytes and inhibited the process. If this turned out to be the case, this "something," Anna suspected, would probably prove to be connected with the iron-binding proteins.

As it happened, Anna learned something else significant about transferrin in Basle. A fundamental protein for cell growth, transferrin is involved particularly in bone-marrow cell growth, and, she now learned (a colleague referred her to a major review on the subject), that the T-lymphocyte is known to have a receptor for it.

In Poland she joined up with Sebastian and, in the train between Krakow and Warsaw, told him what she was thinking and her speculations. In return, he reported that in *his* experiments he had trouble in getting lymphocytes to adhere if there was iron around the solutions. Then, drawing on his remarkable knowledge of cell biology, he filled her in on many of the metabolic cycles in living systems. She learned that porphyrins are present in very primitive systems; moreover, the chlorophyll pathways in plants are very similar to the hemoglobin pathways in animals, indeed identical in some respects, except that in the plants magnesium is involved in the cycle, whereas in animals, iron is. All in all, he was helpful though cautious, encouraging but critical, and he was to remain that way for a long time.

In England, on Sunday, April 24, 1977, I was sitting lazily in bed, drinking tea and reading the *Sunday Times*. Suddenly a headline caught my eye: POLLUTED BROOK COULD HOLD THE CLUE TO CANCER. The article, written by Oliver Gillie, their medical correspondent, reported nine cases of Hodgkin's disease in a housing development near Sheffield. In this one location the incidence of the disease was fifteen times greater than in the rest of Sheffield as a whole, a fact to which the local medical group were alerted when their receptionist noticed that five residents had come down with Hodgkin's disease within two or three years. Follow-up investigations led to the finding of another four cases in the same area, and also one case of a leukemia which seemed related to Hodgkin's disease.

I immediately telephoned Sheffield and spoke to an old friend, a professor of medical engineering, and over the next two days I began to pull in a few more details. The community was in the Ecclesfield area and housed 2,800 people; the nine cases included a milkman, a clerk, an electrician, a laborer, a housewife, a computer programmer, a canteen assistant, a driving instructor, and an excavator driver; not one of the nine had, so far as could be determined, ever met any of the others; there was thus no evidence of a direct infection between them, even though a viral connection was again immediately suspected. (Viruses had been one of the earlier Hodgkin's disease red herrings.) The wet spring of 1969 had been followed by the very dry summer of 1970, and most of the people who came down with the disease had arrived in the area during or shortly after that time. The authorities were looking at the problem of sewage disposal. This had recently been found to be faulty, and in wet weather the sewage would spill over into the brook that fed the local water-supply. Furthermore, sewage from a neighboring farm collected in a pool through which the brook ran. Another possibility was that mosquitoes or midges that bred in the pool might somehow carry the causal agent for the disease. No one had yet thought to look into the possibilities of excess iron in the water, although Sheffield is an industrial town with iron and steel works.

Anna was already considering the possibility that certain people might be unable to handle excess iron through some kind of genetic or metabolic difference, and this disability might leave them prone to Hodgkin's disease or leukemia. It thus seemed worthwhile trying to get hold of her and suggest that she pause on her way home and go up to Sheffield.

Having phoned Glasgow and spoken to Sebastian's wife, who gave me the address of the Polish Academy of Science, I flung off a cable of some length. Three days later the British Post Office reported regretfully that they had been unable to deliver the cable because Warsaw insisted that no such organization existed. I protested. Would they please try again? Two days later they reported that they had again been unsuccessful but I would have to pay for the cable! Clearly this problem would have to wait.

Anna's British Airways letter to Nozaki initiated the transition to a more precise scientific phase in the Hodgkin's disease project. By then the group had accumulated a vast, sometimes bewildering, number of facts, experimental observations, and clues, based on the study of the tissues sent to their laboratory. With a theoretical shape beginning to take on firm outlines in Anna's mind, the possibility now arose of testing her ideas experimentally, and her letter to Nozaki was the first practical move in that direction. When he received it, he said to the others, "This is a very severe letter," but they were not certain what he meant by the word "severe." Later he tried to explain. "It was not a love letter," he said, which it most certainly wasn't. Later still, in the fall, when he was presenting his results to the group, he had another try and this time got it exactly: "It came from Charlie's Angels." A combination of six hundred miles and thirty-five thousand feet, with explicit and unambiguous instructions from someone he hardly knew, had made a lasting impression.

He had got right on with the work, and in the months that followed he was to study the effects of various substances on the phenomenon of "making sunflowers." Very soon he could indeed manipulate this easily, with iron or iron-binding proteins. In addition he was detecting the presence of these proteins in the tissue sections and blood samples, using fluorescent antisera specific to each. At the same time the usual quantity of samples—cells, tissues, and blood—

was still pouring into the lab, and Nozaki was freezing and staining these too. Now every single piece of frozen tissue was subject to seven separate tests, with the lymphocyte suspensions given other tests in addition. They were all immersed in blood, stains, and media.

Anna had returned from Warsaw full of enthusiasm for the iron connection and was deep in thought. In one week, she devoured four fat books, including *Porphyrin and Metallic Proteins, Human Porphyrins,* and *The Discovery of Cytochrome Oxidase.* She ranged through reams of paper—books, reviews, and articles—covering iron cycles, porphyrin and disease, metals, nutrition, enzymes in respiration—anything and everything she could lay her hands on. She found herself led into the bacterial literature and learned that it has been known for some time that certain bacteria need iron both to survive and to become infectious. She read well over one hundred papers on metals and iron specifically, and of all these she was to resonate with five; they provided two or three observations deeply significant for her problem and one or two experiments in iron metabolism which she was later to consider clinching. There were also two or three similarly crucial experiments from the bacterial literature. Privately certain that she was about to strike a rich vein, she absorbed the review recommended in Basle, and there she learned more about the third protein, lactoferrin, about which little was known except that it existed and had a high avidity for iron. Though the binding of transferrin to iron is very high, the binding of lactoferrin is 300 times higher.

Thus, while the laboratory embarked again on a hectic and exhausting six-week experimental and technical blitz, Anna embarked on something different, trying to understand biochemically all about the metabolism of iron and how lymphocytes might fit into it. It was the "thinking time" before experiments—before, that is, she could make precise predictions and set in train the experiments that might confirm them. She was working up to what was to be the major climax of this particular project and, perhaps, of her scientific life.

Yet all was not rosy. Plenty of people were still totally skeptical, and her grant proposals were still being consistently turned down. In the middle of May she had three rejected in two weeks: two by the American Cancer Society and one by the NIH, for which she had been "site visited" when she first arrived.

The turnaround came through a young boy with leukemia whose name was Jesus. One day his pediatrician dropped by the lab to ask Anna if she would like some of the blood sample just taken. (The child had so far had neither blood transfusions nor other treatment.) Naturally she agreed, and the iron stains in this blood sample proved intriguing and significant. Iron deposits were actually *attached to the surface of the lymphocytes*. Anna sat down to think, then came up with a prediction: If she were right about lymphocytes and iron, the serum iron and the iron-binding capacity of the child's blood would be exceptionally high.

Her argument and reasoning went as follows: Nozaki's experiments had turned out as she expected. He could inhibit E-rosette formation—making sunflowers—by pretreatment with iron. Now, she argued, suppose lymphocytes circulate because they somehow "sense" metals; one could then inhibit this circulation by pretreating them with iron, after which they could be covered with the metal and hence unable to detect more of it. She had set two postgraduate students to reproducing this experimentally in both tissues and tubes. Sure enough, the lymphocytes in their tests appeared to move toward iron, and in other experiments, after pretreatment with iron, they stayed where they were—in the bloodstream.

However, in the type of leukemia Jesus had, lymphocytes stayed in the blood anyway. Hence Anna inverted the equation. She surmised that the amount of iron in the blood serum must be very high and would therefore be holding the lymphocytes in the blood, preventing their circulation. She persuaded the child's pediatrician to measure these levels, and her prediction was right.

Normally only 30 percent of transferrin in the blood has iron bound to it. Some 70 percent is always available to pick up and transport any other iron that may come in. But in Jesus's case the transferrin was fully saturated with iron and at the same time lymphocytes stayed in large numbers in the blood. The leukemic child was the living counterpart to the laboratory experiment.

CHAPTER 10

CONCEPTION

Science is the meeting place of two kinds of poetry: the poetry of thought and the poetry of action.

GEORGE AGOSTINHO DA SILVA (in conversation with author, Lisbon, 1978)

The laboratory experiments were yielding gold, and in the four weeks that followed, Anna was either to write to me or record on cassettes the most momentous communications in our whole series.

May 1, 1977

The last two weeks were good. Spent two days in Basle, arguing the case for the cell interaction theory of the control of traffic. Poland was Haydn. The first series of events that have really transformed my past into a distant, still painful echo. The talks I gave were good, but the most important thing was the perspective I gained into myself as the woman I now am and into the exact significance of our finds in an evolutionary perspective. So I am in "absolute peace." I am now going into the lab, and then to the library—there's a lot to be read about plants.

The ecstasy comes from the realization that the process of growth—whether a cancer or a benign growth—is an ancestral one, fallen from meteorites.* I started reading Evelyn Hutchinson's The Ecological Theater and the Evolutionary Play last night. He writes, "The importance of the reduced environment in permitting solution of phosphates in the presence of the normal geochemical excess of iron has not been adequately appreciated." So!

---

*The Egyptians referred to iron as the metal that comes from the sky, having found it in meteorites.

(Tape recording)

It is late. I think I'm still going to start writing a paper tonight. I'm very alone, that is really the most dominant feeling. Very alone. Nozaki is a marvelous boy; Michael is a golden boy; the pediatrician in charge of the leukemias—the acute leukemias—is going to be an important person. She is sufficiently crazy to have accepted what I am saying. It is going to be a very exciting time to live in, but the main feeling is one of tremendous loneliness. It is like walking literally into one of those landscapes of Dali, where the clocks just drip like hanging linen so that you don't even feel the touch of the wind. Everything is still: time is still, the landscape is still, and one feels very alone. It is interesting that the exciting moments are not these. At the end it is all very lonely. I honestly don't know what to tell you. I am tape-recording silence because the silence is very real, and very measurable in time and space.

I came back to the tape again because I sense this is the moment you would have wanted to be here most of all. For here is the moment when I know what I would hope to know—about how lymphocytes migrate, why they migrate, what the things are that make them divide. I am talking about T-lymphocytes. What the things are that make them secrete something. What the consequences are of that secretion. It all comes together in about an hour of fluorescent microscopy. The experiment today was this: Nozaki exposed T-lymphocytes to antitransferrin and then read the E-rosette formation. It is clear that the exposure to the antitransferrin stimulated the cells to produce transferrin. The actual images are beautiful. The sheep red cells take up this transferrin as if they were drinking it. But we did not know this was going to happen; we had done the experiment expecting absolutely the reverse—expecting that the antitransferrin would definitively inhibit the rosette formation. It did not. It just stimulated more production of transferrin. So there can be no question that T-lymphocytes are capable of synthesizing transferrin, and that in itself is very exciting and important.

With this new knowledge, with the leukemia blood of Jesus, and at the same time with the knowledge that yet another grant application has been turned down, everything comes together. I don't know what kind of images come to mind, but they are all walking images.

You know it is like walking along roads. The points are mathematical. Here is infinity, the geometrical point where many parallel lines supposedly come together. Here is the scientific point where many different things fuse. And there, at the center of fusion, it is quite still. People look all around, and because the "thing" could be anywhere and new things have not yet been seen, this is a still moment in time—absolutely still. Every dimension is empty and lonely.

The strength has to come from within, the strength to be still within time and space and not crack up with the surrounding noise and pressure. You have to know that what is important is so important. Therefore one must not be sensitive to anything else. You know it just goes back to the description of the God Within.* One has to have a strength which has to be of an almost divine nature, because one has to be totally independent of any other stimuli—self-beginning and self-ending. One has to believe that what one sees is true, and of course it was true in your mind before it was true in your microscope. You invented it, because you would not have looked for it if you had not invented it within yourself. So even that process itself comes from a completely enclosed circuit.

At these moments perhaps I'm not a human being. It is not human to be so totally self-propelled that you become insensitive to outside stimuli. Clearly I'm still sensitive to positive stimuli. I am still susceptible to joy, but I foresee the day when I shall not, perhaps, be susceptible even to that. The anticipation that my excitement may not be fulfilled is so strong that maybe I shall not react even to joy if it be fulfilled. And that is close to madness, close to the mad guy who sits in there in a world totally of his own invention, populated—for those with megalomanias—by nonexistent generals and kings.

If you want to know what creativity is, it is close to madness. It is close to a level of self-propelled invention, bearing no relationship to ongoing reality. But I know I'm not mad in the sense that no one will—or should—put me in an asylum or in a mental hospital. But the very fact that grants are continually turned down—I've been here nearly a year and not had a single one—means I'm speaking the language that only psychiatrists would probably listen to, and then not out of intelligent understanding but out of professional compassion. Of course I'm very fortunate to be able to play the piano, and fortu-

---

*The name the Greeks gave to the power of man's mind to reach out and understand the world in which we live.

nate to be able to write poetry, because they are tremendous releasing forces which keep one sane.

I expect this tells you enough how I feel. The reason I could not talk before was because there was no space available. The whole of my mental space was occupied with thinking out the theoretical alternatives, reading articles, devouring papers of what other people already know; and that occupancy is of such a nature that one doesn't know how one is feeling. Indeed, one is not feeling. One is totally, totally absorbed by the actual thinking process.

And now that one has time to feel—now that one has time to see that what one thought was right—and now that one is totally alone in that rightness, it all feels very lonely. Didn't you tell me about some man or other* who said that the terrible thing was to have to go and share it with other people, because this is a possessive moment? I don't feel that way at all. I am perhaps so detached from myself that I don't care. I don't think that what is there under the microscope relates to me at all. What is there relates to itself, as people relate to themselves. I think we are all alone, each thing and each person in his own truth. What is truth? The truth is to be a compact molecular complex that, in our case, goes around with legs and in the gulls' case goes around with wings. This living complex is mysterious; it relates to another; it hurts another; it makes others rejoice. It is amazing how such interactions can occur at all.

Now it is very late, close to one or two in the morning. I somehow feel that I didn't want to write all this down. This last two or three days have probably been the most important days of my life. Because not only have all roads of thought come to a point from which other new things have not yet sprung, but, in a kind of way, many roads have come to an end with my life as a woman too.

May 23, 1977

Here I am at the end of a very, very long journey, at the end of a quite extraordinary twenty-four hours. As you remember, on the plane I wrote that letter to Nozaki telling him to do some experiments, and then I went to Switzerland, where I heard something which suggested my prediction might be right. Then I went to Po-

*Albert Szent-Gyorgyi; see Chapter 3.

land, and I talked a great deal and at great length with Sebastian, who no doubt will be doing some experiments in Glasgow. I came back, and throughout this time lots of things happened in me as a scientist, focusing more and more the essence of the problem I have.

Essentially it comes down to something very simple: that lymphocytes—particularly T-lymphocytes—have the capacity to carry on their surface a protein, a metal-carrying protein, called transferrin. The best-known transferrins are those that carry iron. Once they are present, the lymphocytes are able to sense the existence of iron in the body. So, I believe, probably the reason for lymphocyte circulation lies in a capacity to sense iron. We have done some experiments on smothering lymphocytes with iron, and when this is done they don't go anywhere. They all stay in the lungs and the liver.

Then also I thought that perhaps this same reason probably explains why we can even detect T-lymphocytes. So we have been doing some experiments with Nozaki on that test. People discovered that T-lymphocytes make spontaneous rosettes with erythrocytes from sheep, but nobody really wondered why. But I think we now know. All the experiments—every single experiment we have been doing—has been working, and all are pointing in that one direction. So here it is, a fantastic system of sensing metals. It is something that makes great sense in evolutionary terms, to have a system to detect the presence of metals. You see, when metals reach a cell directly, they are very toxic and damaging. But, of course, if you develop a system that perceives and then receives metal, that really improves matters very much.

So then, if you believe that lymphocytes are capable of detecting iron and then binding iron, of course you can understand why lymphocytes play such an important role in infection. Because bacteria need iron to survive! And if there is a system that competes for the very iron that would help the bacteria to grow, then you can understand how a lymphocyte is both cytotoxic and antibacterial!

Also now that Nara has virtually finished—he will be leaving at the end of this week—we find we have the most beautiful study of the leprosy infection, and the areas where the microbacteria grow are, I think, mostly iron-rich areas. That is where the things grow. So it is really very exciting. I don't mean exciting. I'm tired. I have had a very unnatural life since I came back. I have really been spending twenty-four hours a day just thinking, even dreaming of things. I've been in the library, and I have been reading at home, and I've been just thinking. It has been a very unnatural life.

However, the fundamental question is: If all this is true, then one could argue that in leukemia a situation occurs where you have increased exposure to one type of iron—ferric iron before it is reduced to ferrous iron. So you can expect that in T-cell leukemias, especially, you would find a lot of iron in the blood. That was my prediction, and today it came true. It is there. There are lots of red cells showing ferric iron on the surface and, of course, lots of leukemia lymphocytes. I think—I think at least to my satisfaction—that we have made a significant step forward in understanding the mechanism of leukemia. I don't suppose anybody else will believe me, but I do. I am just very moved. At the actual moment where I saw the blood of this child, Jesus, who has acute leukemia, it turned out they had all gone out to coffee somewhere, and there was no one in the lab. So for five minutes I just sat there and looked at the blooming thing and literally cried.

How much detail do you want? I am sorry. I have tried to keep track of myself and it is just impossible. An idea changes, can change four times a day. You wake up thinking it is that, and you go and do the experiment, and the experiment shows that it is something slightly different. It is just like playing tennis—it is like Wimbledon. You go and play tennis with your own ideas but still you cannot keep track even if you live within yourself. This I have tried to do, sitting on my shoulder looking at my own initial thoughts. But you cannot keep track. It goes so fast and changes so rapidly and what is—what has been—wonderful is this: You know how I tend to read widely before doing experiments? Well, we have postulated the existence of something in the morning and then gone and found in some remote journal that somebody has in fact found evidence for the idea. And it just has gone on like that. In fact, when you put the existing evidence together, and you write what you are thinking, then it all comes together. But don't ask me how it happens, because all I know is that it requires an absolutely twenty-four-hour, sixty-seconds-per-minute, sixty-minutes-per hour type of concentration. Everything else becomes peripheral. People become peripheral. Food becomes peripheral.

After receiving this tape, I telephoned. Anna told me that Tien-

Chun, with true Chinese metaphor, called her a kite. She herself was hanging on for dear life to be pulled as Anna flew. "If I go on hanging long enough," she had said, "I will understand." Anna kept saying that she was doing just what she had come to America to do. Her experiments had worked beyond expectation. So far as she was concerned, the answer to the question: Why do the lymphocytes circulate? turned out to be also the answer to why children get leukemia. Iron would prove to be the beckoning siren. Finally she said, "You know I've spoken a great deal about making love and making understanding. Well, now I really know the difference. Love is light, warmth, and comfort. Understanding is just light."

"I could never be a scientist," I told her.

"Why not?"

"Because I don't have the inner strength. I need warmth."

"I am glad you said that. I'm glad you recognize the problem for what it is."

The letters kept coming.

May 30, 1977

I sat by the window as the river flows and your letters miraculously appeared in the letter box. I cried at the waste of Avivah's death and your crying for Avivah's death.* Above all, I cried for the stupidity of my own desire for clarity and understanding. Because at this stage feelings of this sort are premature. On April 29 I went back to all the books I had bought in Chicago. I started eating the words right away. The days were not things of light and a few hours, but total lumps of time through which I went from one idea to another, to one experiment to one dream to another idea. When in your letter you said that it is only three weeks since I got back from Warsaw, it shook me to think that there were such things as weeks. I had forgotten the divisions of time. I had spent my days in the laboratory; the nights in the library and part in bed, dreaming of oxidation and reduction of iron. There was a total sense of urgency.

Things in the lab started to come to a grand climax, like an enormous snowball. No, like going on a mystery tour in the tunnel at a

---

* See dedication: A.Z. died of cancer, as did the other two friends to whom this book is dedicated.

country fair, and finally seeing the minute spot of light. We went from one experiment to another to another to another. All working. I've had no time for relationships other than close ones. Whether these go on to be expressed as love or just as blessing and binding friendship is immaterial. Both for me are a natural outcome of being alone, and I have no time for superficiality. No time.

For iron dots sit there under the microscope, in the peripheral-blood red cells of the child with acute leukemia. You see the urgency of my need to clarify the iron story. Everything else is blissful bonus.

Sebastian is now getting exciting results with iron.

At the same time, a conflict between her personal needs—the lamb in her—and the demands of the scientific lion was adding to Anna's tension. During one transatlantic telephone conversation I asked if she was ever tempted to leave it all, let it all go and become a lamb again. Could she do that?

"I doubt it," she replied. "Don't you realize there are twenty papers to write? No one else has seen this; I must do it. And as far as your question is concerned, the answer is I'll do it when I'm dead. That's my epitaph. I've been a lion all these years, wanting to be a lamb, and there of course lie both the ecstasy and the misery.

"Besides, my instinct tells me about this love for things—this special relationship for things—tells me that I may not manage to cope with people. The instinct is the inner necessity, at once compelling and impelling. In any case, I don't think I have any alternative. It is difficult. I suspect I'm almost too direct with people. I have almost too much clarity, and they can't cope with it. It has taken so many deaths for me to arrive at this point. I am a young face in a young body in a young form, laughing and happy all the time. But really I'm an old woman inside, old from so many experiences. I still resonate to the purity and the clarity. I'm a mathematical formula, and I long to die. Mine is the eternal problem of women in science. Very, very few women bring this off. But it is our own choice; it is self-inflicted."

This is, of course, the eternal problem of *anyone* in science. It yields its secrets only to single-minded obsession, and at the end of the lonely day, the week, the year, or the decade one emerges from the laboratory into what, humanly speaking, is a very empty place.

May 30, 1977

I am exhausted. I want to cry endlessly, the lion, the lamb, and the mathematical formula. You have no idea. This last month and a half has been the most creative, the most productive, and the most rewarding of my life.

I am just writing paper after paper after paper. From one single thought we can now interpret and unearth data in the field of infection, in the field of leukemia, in the field of physiology. It is like the first rain after the drought. The suddenness of it all is a bit overwhelming. To describe it to you I have only one word: clarity. The clarity of whitewashed houses and the clear blue skies in St. Barbara de Nexe. The sense of ultimate clarity that once again brings me very close to a sense of death, of becoming the ultimate chimera. You put it beautifully. My death will be just that: the final merging of lion and lamb together in one mathematical peace without wind.

Tuesday (perhaps)

At the end of a crowded day, two other papers virtually out of the way, and here I am again back facing an empty yellow page.

It is immense pleasure to work with Tien-Chun. We really enjoy working together; we are sufficiently well established to enjoy, without fear, the collaboration.

I must sit down and get at least the framework of another paper out tonight. Today the boys, Michael and Nozaki, actually visualized what is perhaps the essence of lymphocyte traffic—lymphocytes moving toward complexes of iron and transferrin. Really exciting. Michael in his understating fashion just said, "We are having fun." Nozaki smiled from ear to ear, saying, "It is marvelous."

June 10, 1977

I am so tired of so much emotion that I can hardly gather the strength to write. But I must tell you that the little boy, Jesus, with acute T-cell leukemia and iron spots in his red cells, has an iron level in his blood serum of 456 when the normal range is between 48 and

193. I must therefore tell you that from little mice we have traveled all the way to where it matters—to little boys. .

From little experimental models made of history and past ideas, we have traveled all the way to the beginning of dawn, where from a dark mass the day begins to get shape and promise. And it is a cold hour.

And—as I said—that is what makes it different from love. For love is light and warmth, and understanding is only light: only much later, and dissociated, is it warmth.

I am very tired and very hungry and in a kind of way closer to Nozaki, Michael, and Carolyn than to anybody else alive. I don't know whether they know it or feel it, and I am not going to tell them.

So after a long hot bath, I shall go to bed. From now on, it will be repetitive and confirmative, until the other important, important time, when sufficient and substantial evidence will be available for enough conviction, that will then change our present ineffective cancer therapy.

June 23, 1977

CERTAIN TRUTHS: Not only are we sure of the association of iron with T-lymphocytes, but the cells we looked at today may well tell us the nature of the malignant cell in Hodgkin's disease, and it really may contain iron.

June 24, 1977

I wish you were here. Time again for simple things. I have kept my sanity by playing Bach. Bach is wonderful, the very clarity and transparency of the mathematical construction makes it a tremendous releasing experience from all the complex things that are going on. But so do the kids. Michael is golden and so too is Carolyn, blossoming in her own experiments. So too is Nozaki in his oriental humor. I am afraid I have very little else to tell you. We had a most delightful first birthday picnic for Juliette [Michael's daughter] on the Roosevelt Island grass.

July 5, 1977

Work: Michael has just phoned with some results and you will see later why, when I told him, "You are keeping me alive," I was speaking the truth. Message from a colleague. Went to see him. He read the leprosy paper that Nara had written. I had changed very little—deliberately. He wanted to talk about it. If I am to be part of his group, this is a test case. And the test is whether one settles for mediocrity or not. You know my answer to that.

Mediocrity is the thing that gets published and gets easy grants. But we are going to do work that takes a very long time to plan. To do the new experiments my colleague suggests will probably take another six to nine months to complete and to analyze to our satisfaction. So there can be no working within deadlines. There just cannot. But there can be no falling in love, either. There is no time or space in this city for emotion that paralyzes, for emotion that hurts. No time for ecstasy, just not time or space in this city.

As I sit here in the living room there grows in me a poem which will be called "The Condemned." I have the alternative of the window or the door into the room where I am working. And I choose the room, but there is no glamour in this choice. It is all hard work and the way to duty. The woman in me wants to give up, but the total person I am is here, and rejoices in Michael's voice telling me that ferritin-positive and transferrin-positive cells are in the thymus-dependent area of this lymphoma spleen, where macrophages loaded with iron are also present. I rejoice for the finding and for Michael's voice. I told you that without their knowing, I'm closer to them than anyone else in the world. As it is also true that they are keeping me alive.

As for what I did in May, I hardly know. But by the end of the week—May 22 to be precise—all predictions were coming true. But I hardly knew my own daily life. Last week when I collapsed in bed on Sunday at four o'clock in the afternoon, I realized that I hadn't had a meal at a table since Tuesday, and this is probably true of most of the days since I came back from Poland. I have had bouts of thinking. At the beginning I spent most of the time reading because I needed the background knowledge I did not have. Afterward I just lay for hours on end, just thinking. I've done very little beyond reading, thinking,

*playing the piano, and at the end of May writing some reasonable po-
etry.*

July 9, 1977

*I finished writing one paper and am just about starting to struc-
ture the beginning of the other in the series, on the lymphoma
spleens. Yesterday's lymphomas—the spleen sections—had the thymus-dependent area absolutely choked with iron, impossible for even
the slimmest lymphocyte to pass through.*

So the spleen becomes bigger and bigger and BIGGER, *full of stuck
lymphocytes.*

In July 1977 Anna took a quick trip to Madeira for a meeting of
the Portuguese Society of Pathology. En route, in Lisbon, she looked
at sections from one case of Hodgkin's disease which had already
been prepared and stained for her by Dr. Maria Eugenia Horta,* a
pathologist and longtime devoted friend. The sections were full of
iron.

At the meeting she gave a paper to old friends about new findings. She also gave Dr. Horta details of what to look for and asked
her to see if she could confirm the disease: to see whether in her
Hodgkin's disease patients the crucial identifying Reed-Sternberg
cells had iron in the cytoplasm. She also requested sections of any
normal spleen they received.

When she was back in America, we spoke on the phone.

August 1977

*As you detected well on the telephone there was something
wrong, a very fundamental form of nostalgia I cannot describe. I'm
going through feeling under-the-bed-size small—small, small, ready
to hide in any small space you care to mention. I decided to stay
overnight in the lab tonight. Do not feel sympathetic for anything.
I've just been dying for long moments of reflection and silence, and I
generally do not have these in Portugal. So here I am at the begin-*

*See dedication: M.E.H.

ning of another night of metallic silence. I'm scared . . . scared stiff, of the repercussions of our findings. I phoned Maria Eugenia Horta this evening and she has confirmed that some of the Reed-Sternberg cells in her cases of Hodgkin's disease do have iron in the cytoplasm. I had spent the Saturday in Lisbon looking at her material and showing her mine. The meeting in Madeira was important because I tested the findings and the ideas against an audience of very knowledgeable, very traditional pathologists.

And their reaction was not "What nonsense" but "Be cautious and continue, and we will look at our material." There is the root of my nostalgia—their support. The confirmation today no longer made me jump, or feel excited, but nostalgic: nostalgic for all the things of peace and simplicity that will vanish with yet another step forward. It is very strange but there it is. You are probably a bit lost and hardly know what I am talking about. Let me sum up.

A. A central defect in the macrophage. A defect in the handling of iron in the macrophage could explain many well-known empirical observations about Hodgkin's disease: the low serum iron, the high iron storage, the slow release of iron; the trapping of the lymphocytes, the failure of cell-mediated immunity and the normal spleen responses; iron deposits in the lymph nodes; the Reed-Sternberg cell; multiple cases of the disease that might be associated with water pollution or too high an iron intake.

B. Lymphocyte circulation. If the basis of the lymphocyte maldistribution is the lymphocyte's ability to detect the presence of iron, or of iron-binding proteins, we can easily test the hypothesis that this is what lymphocytes do normally by pretreating them with metal irons. So we have, and the results are very encouraging. Lymphocytes pretreated in vitro with ferric citrate, for instance, either stay in the blood and go nowhere or go as far as the lungs and are delayed therein. Terribly simple and it works.

C. The mysterious T-cell receptor for sheep erythrocytes. Human T-lymphocytes make E-rosettes (sunflowers) mysteriously when mixed with sheep erythrocytes. We worked on the hypothesis that the T-lymphocyte "sees" metal on the erythrocyte, and we did the same experiment as before. Lymphocytes pretreated with iron salts (as opposed to those treated with zinc) do not form rosettes. The more extensive aspects of this study are still going on, but it is all pointing in the right direction.

Nozaki's older daughter was hit by a car today and has her little leg in plaster. The sense of being responsible for these people also makes me nostalgic—nostalgic for a state of irresponsibility.

Do you understand this aspect of scientific process? One is no longer proving one is young and clever. One is old and clever, and responsible both for those people around and for the things discovered. So I miss watching the cricket Test Match. Something as simple as taking time off to watch the Test Match is incompatible with the inner rate of discovery at the point where the facts are known only in your own lab. It is even more incompatible with the doubt that it might be all insignificant and explained in some other way that you cannot see.

So I yearn for the day when I shall watch a Test Match again without any discovery in the works!

The summer progressed. Though grant proposals were making no headway at all, though the collective group chant was the despairing cry "Scientifically approved but not funded," there was to their relief evidence of mounting interest in the work. A whole series of related events occurred in mid-June. Friends made substantial financial contributions; a paper appeared in *The Lancet*, the British weekly medical journal, about the mystery of the multiple Hodgkin's disease cases in Sheffield, and the story found its way to America; earlier in the year New York newspapers had reported that certain wells in New Jersey had been closed, partly because of contamination but mostly because of too high an iron content in the water. An outbreak of Hodgkin's disease and leukemia in the affected area had caused an immediate investigation into possible causes, and Anna—together with her hypothesis—had been drawn into it. She and Tien-Chun had constructed a questionnaire for the patients asking about their swimming and water-drinking habits. The answers indicated that many of them lived near, or had access to, well water with a high iron content. Others seemed to be "living in the swimming pool," and chlorine, of course, releases iron from water. More and more isolated scientific papers turned up, some old, some new, all with some bearing on the problem, and the group's researches into the literature began to reveal more links between the various metallic proteins and the lymphocytes; evidence slowly mounted that lymphocytes do interact with ferritin and transferrin and, perhaps most importantly, with lactoferrin.

As Michael later confessed, their linking of the iron metabolism with the immune system was thoroughly disturbing—and a very good thing, too. Scientists fall too easily into the habit of assuming that the different aspects of biology are unconnected, believing for example that iron cycles are one thing and the immune system quite another. But as it became more apparent that the lymphocytes were intricately involved with certain metals in the body, Anna was prepared, on the evidence, to make an imaginative jump and connect the two.

They received some slight encouragement from other people within the institute, though by no means total acceptance. The head of the lymphoma group, a scientist universally admired for his fairness and competence, met with Anna, discussed the work, and promised a measure of understanding support. Finally two test cases gave the group the opportunity to try out all the techniques they had developed within the framework of Anna's working hypothesis.

Two Hodgkin's disease patients underwent surgery, and samples of the spleens came down to the laboratory. One, in Stage I, had no obvious tumor nodules; the other, in Stage IV, was extensively involved with the disease. For a while there was organized pandemonium while everyone went to work. The results looked good and seemed to confirm their earlier tentative findings. They were also able to follow the progression through monitoring the two patients. As the disease spread, increasing numbers of cells having the iron-binding proteins appeared in the tissues, and more cells containing iron appeared in selective areas of the tumor. The cell suspensions also showed clumps of lymphocytes staining positively for iron and the three proteins.

By the end of summer the group could even sense some interest from the professional medical staff. So Michael, working frantically, was one day moved to say, "If we're wrong, we'll all go to jail. But if we're right . . ."

We waited. A medal, perhaps? A grant? That would be nice. ". . . Perhaps someone will take us out to dinner."

At the end of August Anna took a short holiday, and she and I finally met up again in Oxford, at a conference on Cell Interactions organized by Sebastian. There, in the vast Victorian halls of the old

Museum of Zoology at Oxford, enveloped in dank misty days and sufficient wet cold to remind us that Oxford sits unhealthily between two rivers, I met a number of the characters in the story, among them two scientist friends with whom Anna had been professionally involved for almost fourteen years. The conference was very much a mixed bag, but two or three events stand out, as did one particular gem of a paper on leukemia by her old friend, Francis.

Early in the conference Anna presented some of her new findings and tried out her new hypothesis. She did it badly. This often happens when her enthusiasms and thoughts propel her far ahead, and she takes it for granted that the steps in her argument are either as logical or as obvious to others as they appear to her. More often than not they aren't, but still she tends to leave them out. Thus though her empirical facts are acknowledged as sound, her argument can appear thoroughly disconnected, unproven, and, also, too showy by half.

Studying the faces of those who had been her immediate colleagues or peers, I saw reflected there the same suspicious pain as when Nara took the live mouse into the microbiology department. They either didn't believe a word of it or they didn't *want* to believe a word of it. Here, one could sense them feeling, was this "madwoman" going off yet again, this time making a totally wild suggestion about the association of lymphocytes with iron. At the same time, one could also sense that something held them back from blanket condemnation. They were just faintly, nigglingly suspicious that she might be on to something after all; they dared not dismiss the idea outright because she had pulled something off once before and might just be about to do it again. Still her reception was very rough indeed. It was Francis, with whom Anna gets on splendidly, who asked the crucial question: "What is the evidence for there being transferrin receptors on the lymphocytes?" Anna gave *an* answer; at that stage there was some evidence but not very much, and it certainly wasn't clinching.

Later on, Sebastian talked privately to me about the problem. He was quite prepared to believe that Anna was on to something very important and could well be right. But he was extremely worried as to whether her phenomena were real. For as many others had pointed out, one real possibility was that the pretreatment of the iron was exposing the lymphocytes to substances of such toxicity that they were all dying. In that case their failure to react normally in the "sunflower" tests could be due *not* to the blindfolding of the receptors responsible for such reactions but to the fact that the cells themselves were, if not dead, at least thoroughly poisoned. In addition, someone had to see whether the iron stains were arranged in some sort of pattern on the surface of the lymphocyte. If there really *were* specific iron receptors on these cells, one would see the clumps of blue stain very precisely organized over the cell surface. But if the stain was just plastered on all over the place, with no discernible pattern, it was very likely that the iron was clinging to various areas for some indeterminate reason and not attaching itself to specific receptors. Anna felt the force of this criticism acutely, and she determined to send the lymphocytes, both pretreated with iron and untreated, to the electron microscopist for a whole series of tests.

CHAPTER 11

BUILDING

I'll huff and I'll puff and I'll blow your house down.
BIG BAD WOLF TO THE THREE LITTLE PIGS

Anna is subject to a recurrent panic: that she returns from vacation only to find that she has forgotten *everything*. I returned from Europe with the same worry, and so in the second week of September 1977 I made a quick tour of the laboratories, taking about an hour of each person's time and catching up with the work. Carolyn was in her usual place, under the laminar flow hood, and, as I might have guessed, she was working away at spleens. She talked while she set up the cell cultures for a spleen and lymph node that had just come in, the practical activity providing a counterpoint to the themes of the summer as she recapitulated them.

"I always need a minimum of ten grams," she said. "But the more the better. Not more than ten hours old and hopefully much less. I have to keep it sterile. I have to keep it in physiological balance with the proper salt solution, and I have to put in penicillin and streptomycin to prevent any residual bugs from growing in the culture."

To my untutored eye, the spleen looked like a piece of calves' liver, but it had small white bumps on its surface. Carolyn cut off one piece for frozen sections. It would go straight onto the block and into the cryostat, where the sharpest of knives can cut the thinnest of slices. On some sections, staining procedures would be done, others left clear, but all would be examined for fluorescent cells.

"It's a slow, beastly procedure," Carolyn went on, "and the more you have, the more tedious it gets. I work away at it with up to eighteen needles, just teasing, teasing; trying to get the cells away from each other. If you look, you can see the cells float up into the media. I do the first wash three times, and I do a new teasing between each wash. I use a Pasteur pipette to take up the cells."

The Pasteur pipette has a very long, thin end, and even when Carolyn—who is quite a young person in science—began her career, scientists had to make their own. It used to take about one week of glass shaping to make one box of pipettes, and this group used at least one box a week.

"I'm taking the last wash off now. The summer was dreadful— spleens, spleens, spleens. Spleens and the hairdresser, spleens and shopping, spleens and rain; and it was much too hot."

She was now drawing up the cells into an enormous plastic hypodermic syringe to get the exact amount she needed in her vials, which were already prepared with the culture medium. Dropping in the few cubic centimeters of spleen-cell suspension, she made a qualitative judgment: The solution was too dense and she would need to dilute it further to get a good separation of the cells in the centrifuge. I watched her slide the top layer of cells onto the culture medium. It was another slow, tedious part of the procedure, but it looked as if she were making Irish coffee. Then, as the cells started to separate out, the fluid took on the appearance of that mixed drink where the different liqueurs separate into their layers. On top was a beautiful layer of pink; at the bottom a completely transparent layer; in between was a very small interface of opaque fluid, where the lymphocytes would sit after centrifuging. In the centrifuge the red blood cells and the polymorphs from the spleen would spin out into a small pellet and settle down at the bottom of the vial, and the lymphocytes they wanted could be gently pipetted out.

"It's going to be spun for thirty minutes at thirty thousand times the force of gravity, and I have to keep the levels absolutely exact, so that all the vials are balanced in the centrifuge. Otherwise you can screw the whole thing up."

So saying, she went off to do the mitogen studies: "feeding" some of the lymphocytes with those extracts of plants which, earlier scientists had shown, stimulate the cells to divide and thus provide an index of their vitality. She would monitor the divisions and measure them by incorporating a radioactive liquid into the specimens. This would enable her to see how many of the cells in the initial specimen had divided and to what extent. It is not known just *how* the mitogen works to stimulate the cell, but at least the test tells the scientist how competent the cell itself is: whether it is in a position to do anything at all immunologically or whether it is, biologically speaking, a "dead duck."

"I think," said Carolyn, "that this is the most stunning example of the use of mitogens. We use it to look at the viability of the cell populations and at Anna's lymphocyte populations in her ecotaxis experiments. I have to be very gentle. The gentler I am with the cultures—the more time I take—the more viable my cells will be, and remain. So in all it takes an enormous amount of time."

Everyone agreed that during the summer it had been Carolyn who held the lab together. She was the person to whom they turned if they wanted anything practical sorted out. She knew the ins and outs of the institute like the back of her hand and thus carried the detailed burden of everyday existence. The day when she stayed home with flu, her telephone rang twenty-five times in the first two hours, and there was no shielding her.

"Do things feel good to you right now, or is it a little mad?" I asked.

"Fortunately it is both. If it wasn't a little mad I would lose interest. I see other labs plodding along with the same thing and getting nowhere. We seem to be starting to expand outward. I am convinced there is certainly something here. It may not be what Anna initially picked out, but there is a big system out there, waiting to be tapped. It is the feeling of expansion rather than contraction which feels so good. Of course, Anna sees, or feels, the overall picture more than we do. We just see little bits of evidence dotted around, and it is sometimes difficult for me to put it all together. Michael and I often have to sit her down and make her go through all the evidence to pull it together for us. There have been times when I thought the whole thing was complete rubbish and there was absolutely nothing we were looking at. At first, the fluorescent work was too comparative. But now people around here are at last reasonably interested, and some are totally amazed. Others are much more worried for Anna than for the theory. Right now it is very theoretical. But I think she has decided to stick her neck out, for better or worse, and I really admire her for that."

"Why?"

"Because she could be in for a lot of trouble. Yet I'm sure there *is* something there. I think she may be opening a new field, so even if she gets laughed at and slapped down to start with, people will begin developing it and start looking. And sooner or later, people will come up with the really hard, confirming data for what we, at the moment, are just scratching at. But there it is. Anna to me is totally innovative. I've never met a scientist like her. She doesn't come across as a scientist at all because she seems to be so 'unscientific' in her approach. She remains very open-minded about results. She's very willing to be proved wrong, and that is the compensating factor. She has a quest for the truth without being enclosed by it. I find some scientists have very set ideas and will go out to prove them at all costs. They will not bend according to what they see. Anna will look very hard for what she wants to see, but if it is not there she will accept it."

In another laboratory were the two postgraduate students, Joe and Nancy, working toward their doctorates.

"I'm now basically looking at the effect iron has on the traffic of lymphocytes," Joe said. "What I do is really very simple. I have a solution of about three cubic centimeters of ferric citrate, depending upon the concentration of cells. I leave the cells in this for half an hour at thirty-seven degrees centigrade. Then I wash them three or four times. I count them and inject them into mice of the same strain. Then I go look for them. Since the lymphocytes have been tagged with radioactive chromium, I can trace where they go and follow them after one hour, twenty-four hours, and forty-eight hours."

Nancy's role was to look at the organ histologically after Joe had done his experiment. She took the tissues from his animals and examined the cells closely. He was looking at the animals grossly, to see where the cells were *going*. She was doing with mice what Michael was doing with the cells from the human patients, testing them for the presence of iron and the iron-binding proteins.

"Transferrin," Joe explained, "and the iron cycles in the blood have been around for a long time. People have worked out elaborate biochemical schemes for the iron cycles, and these have already been expanded in great detail. What our group is now saying is that these iron-binding proteins have one more important role—an immunological role—and we are going to have a lot of people on top of us trying to knock the work down. It is important for us to develop it strongly at the beginning before they can start."

Among the three iron-binding proteins, lactoferrin was now the group's current preoccupation, and they were concentrating on it hard. It was suddenly to make its presence felt most convincingly. One evening Anna returned in a mood of impotent frustration from visiting a very sick young patient in the hospital. The boy was running a high fever; a mass in the middle of his chest was growing at an alarming rate; he was getting progressively more anemic. Every imaginable test had been done, and nobody had any idea what was wrong. Even a biopsy revealed nothing definite in his cells.

Because of the complexity of the case and the bafflement of the clinicians, virtually every laboratory was involved. Anna's group stained sections of lymph-node tissue for the presence of transferrin, lactoferrin, and iron and thus discovered that the lymph node contained many lactoferrin cells. In fact, the normal architecture of the organ had completely disappeared as it had been invaded by many different types of cells. When they studied the cell suspensions of lymph-node tissue, they saw that the cells were in little clumps, surrounding a central area of material they couldn't quite identify. But these clumps as a whole were staining positively for ferric iron, and some of the cells surrounding them were staining very heavily for lactoferrin. Anna wondered if the patient was producing too much lactoferrin, and if the lactoferrin, always thirsty for iron, was scooping it all up so that none was available for the synthesis of hemoglobin; this would explain the boy's anemia. Or, perhaps, his increasing anemia was due to a lack of the transferrin necessary to carry the iron where it was needed. Perhaps both factors were in play. Samples of the boy's bone marrow and blood were therefore stained for transferrin and lactoferrin. They revealed that 80 to 90 percent of his bone-marrow cells contained lactoferrin, whereas very few contained transferrin. Anna concluded that there might indeed be an abnormal expansion of lactoferrin-producing cells which were then scooping up the iron, but she had no idea why this could have occurred in the first place. She had, in any case, already instituted a routine examination of *all* tissues for the presence of lactoferrin. This routine had been in operation for some three months because she thought they could perhaps thereby find a pointer to explain why some people—rather than all—contracted Hodgkin's disease or leukemia.

At this time Dr. Marion McLeod arrived for a visit, bringing several sections of skin from various patients, including those with mycosis fungoides, a skin cancer in which the lymphocytes are again trapped. She wanted the sections stained for the three metallo proteins. Of course, she also needed controls of normal skin. At such times one knows what is expected, and three days later a core was punched out from the skin of my back and stained for iron-binding proteins. Thus I saw for the first time a pattern in normal skin which was to be revealed over and over again. There is a band in the outer epidermis that stains green; it looks like a section through a jungle and is full of lactoferrin. Indeed, this outer band of lactoferrin lines the skin, the membranes of the mouth and throat, the gut, and, it has been reported, the vagina also. Immediately beneath this, in the dermal layer, is the transferrin.

Tentatively, Anna put the following interpretation on this pattern: Lactoferrin was there as a preventive layer to stop absorption of iron. If you had this protein you were, within reasonable limits, all right so far as iron and excess iron was concerned. Whether you had ingested iron yourself, or were exposed to it, you could handle it. But if you didn't have that boundary layer, you might have problems.

In the first week of October she was greatly encouraged. Samples of normal human spleen were at last coming in to serve as controls, and everything was beginning to make tremendous sense. "Good for God," she said one day. "This is one of the most beautiful biological systems ever invented. It is simple; it is beautiful; it is unbelievable. Bacteria and tumors *both* need iron. I now know why macrophages kill tumors; I think I now know how, as well as why. Macrophages take the iron away, so the tumor is killed because it doesn't have any iron to utilize. You have something in mind; something that exists in your mind only; you predict it. Then it comes true, and you find it really does exist and is probably in the literature too, if you only go and look through the whole of it! Oh, the pleasure of it!" she went on. "The prediction becomes almost redundant; the idea becomes believable."

Her enthusiasm was so infectious that I was moved to quote Newton in romantic reply: "I do not know what I seem to be to others, but to myself I seem to have been like a boy walking on the seashore, picking up another pebble more beautiful than the one before, while the great ocean of truth lay all undiscovered before me."

"It is more like going on a picnic to Blackpool," Anna countered flatly. "And there are all these messy people everywhere with beer cans of irrelevance strewn all over the place. I can't stand people who just put everything together and try to see what happens—who don't *think* before doing an experiment. They upset me. I get the same uncomfortable sensation as when I see a picture hanging crooked on the wall. I could never do science like that."

Then her mood changed, and she became very sober.

"I've never been like this before in my whole scientific life. Perhaps I'm taking it all too seriously, but for the first time I'm really frightened."

"Frightened of what?"

"Frightened of making a mistake, of being wrong, and that is terrible; it is just terrible."

She had—and she knew it—to face the possibility that after all her critics could be right; that though her individual bricks of discovery were sound, the theoretical edifice she was constructing was the wrong one. The possibility was giving her a bad headache, she said, and worse.

"It is a funny feeling. I'm on the verge of explaining so much, but it has taken me by surprise as much as anyone else. It is like having gone through the wrong door and suddenly you're on a big stage, *all alone*. And I ask myself: *What am I doing here?* I feel like a small child. I want to go home!"

Now it was time to compose for the institute the annual report of the section's work. She enjoyed that, taking two whole days to write sixty pages, only to be told by a "sweet, kind" vice-president that it wasn't what they really needed. Two paragraphs were quite enough! Her exasperation shot geyser high. "What kind of system is this?" she asked. "In order to get a grant, I am expected to write sixty pages on what, if I'm honest, I can't possibly know I will be led to do and can't say where the clues will lead and can't anticipate the results. But when I write sixty pages telling them very precisely what we *have* done, where our thoughts have led, what we have achieved, what we have discovered, what we think we have understood, I'm told only two paragraphs are needed. This is a looking-glass world."

But then two events shifted her mood back. Sebastian sent her a birthday card, with word that he could get T-lymphocytes to suck up iron. (Months later, on January 28, 1978, he telephoned to say that he had found iron receptors on B-lymphocytes too.) And a paper came in for her to referee, dealing with tumors and a radioactive substance called gallium-67. For some unknown reason, gallium-67 seeks and latches on to some tumors, and the amount of gallium-67 the tumor cells take up is a test of their degree of malignancy. When blood serum is added to the cultures, the uptake of gallium-67 increases. The paper seemed to indicate that transferrin was the component in the blood responsible for the cells' extra thirst. Anna liked the paper, played with the figures for her own amusement, and came up with a conclusion the authors hadn't mentioned because it didn't enter into their initial inquiry.

"It is interesting," she said. "In the cases cited in this paper, people with Hodgkin's disease are not markedly different from controls in their uptake of gallium. But they are *strikingly* different in their uptake of iron—and even more so when transferrin is added."

"Why?" I asked.

"Oh, my God!" Anna exploded. "You are just like my mother. She is always asking why? *I don't know why!* It is enough that I've demonstrated the fact. I've got this idea about iron and lymphocytes, and that Hodgkin's disease is an abnormality in the handling of iron, out of the blue—literally out of the blue. Now here comes along a nice piece of evidence that I'm probably right. That is *enough* for the moment."

Joe's words about people trying to "knock down" the group's work were to re-echo a few days later, when Francis, Anna's longtime friend who had given the star paper at the Oxford Conference, visited New York. He, the director and his wife, and Anna and I met for dinner.

Since all three scientists were distinguished immunologists, the talk inevitably turned to Anna's new ideas. Their exchanges revealed both the attitudes then dominant in immunology and the characteristic reactions of scientists confronted by new theories. The director began to talk of the simple cures that are possible for certain immunodeficiency diseases in children when they are put on a diet containing zinc. Various metabolic reactions are totally dependent on zinc, and thus such cures can be quite dramatic. This was

all very well, but Francis was reluctant to believe that equally simple principles would apply in other situations.

"I think," he said, "you are being wildly optimistic."

The director reacted rather sharply, with a mixture of irritation
and conviction. "You've got to be optimistic in this game," he said.
"Cancer is such a difficult disease. You can't give up impotently, in
the face of our ignorance."

Was this, I wondered, conviction or just courage? It was both,
actually. This stance has recurred over and over again in the history
of medicine, whenever people have been needled or persecuted for
their ideas. The story of medical progress is the story of continuing
optimism in the face of ignorance and skepticism. In fact, *both* attitudes are necessary. The problem is to strike the right balance.

When the talk turned to iron and lymphocytes, Anna argued
strongly for the possibility that the immune system had evolved
during biological time for reasons beyond the one currently known,
defense, perhaps to cope with the excess iron in the environment
and with iron or other metallic poisoning. She reminded them that
many bacteria need iron to become infectious and toxic; this is deep
in the traditional literature. Thus a system wherein lymphocytes
were perhaps taking up the iron that the bacteria needed would
produce an environment unfavorable for infection. Francis began to
object, not to the facts, which were indisputable, but to the implications.

"You," he said, "are probably saying something very important
clinically about chemical conditions in the body, but I don't believe
you are saying anything about basic immunological function."

"Why not?" I asked. "What about the 'iron' function? Is it not
possible that the immune system has an additional job over and
above the one immunologists now give it?"

His answer was an emphatic no.

"Why not?" I asked again.

"Because in science," he said, "you must keep things simple. You
look for simplicity in your explanations. As it is we already have a
sufficient and simple explanation for immunological function, and
there is no point, and no need, to burden things and complicate theory further."

He was reflecting a reasoned, reasonable, and long-hallowed scientific attitude. Anna stayed silent. The director was half laughing, half exasperated.

"Could *you*," I asked him, "entertain the possibility that the immune system has this additional function?"

He suddenly roared. "Sure I'll entertain anything, but I don't want to be bored! I loved it when Anna explained everything by cells being in the wrong place. I understood it when she said the lymphocytes should be here and they weren't, or they shouldn't be here and they were, so there was a disease. But now she's gone and spoiled it all with this stuff about iron. She's messed it up. It was so simple and beautiful before."

We laughed, and suddenly Francis burst into the well-known chant of the Cockney scrap-metal man: "Any old iron? Any old iron? Any, any, any old iron?" Everyone joined in, but Anna, although laughing, was very grave.

I finally said, "It is all very solitary."

"But that," said the director flatly, "is how the best science gets done."

When the others had gone their ways, Anna turned to me and said simply, "I'm right, all the same."

It was going to be a long, hard winter.

A few days later, Anna began rewriting the paper she had given in Chicago for publication with the other conference papers. "I've decided to put all my eggs into this basket," she said. "I shall stick my neck out about Hodgkin's disease and iron-binding proteins. I rang Tien-Chun and asked if she minds having her name on the paper in view of the risk. Because if she wanted to have it taken off, of course, I'd agree. She didn't. If she had said 'Take it off,' actually I'd have been shattered. All my former colleagues are worried about my doing this. They act as if I need protecting. But if I'm wrong, I'm wrong. In any case, what does it matter? Phenomena are phenomena are phenomena. The proper role of the scientist is to be so humble that you obliterate—eliminate—yourself completely from the picture, and then you begin to understand it. What do *we* matter? Anyway, I can see how the iron messes up the immune surveillance system," she continued. "It blankets it by latching onto the iron receptors on the lymphocytes. But how on earth am I going to prove it? I can also see how a feedback mechanism might inactivate the macrophage. So now I'm going over to try it all out on Jim Hirsch." It was November 1, 1977.

James Hirsch has been a pioneer in research on macrophages and granulocytes, those white cells with branching nuclei that contain granules. I ran into him shortly after Anna had seen him.

"It is very interesting, indeed," he said, "but it is still much too early. It may be a red herring. Maybe lactoferrin doesn't stop the iron from coming in. Maybe it only stops it from getting out. You know, after twenty years, we *still* don't know what the function of lactoferrin is."

"Well," I said, "it's an interesting time to be around. Anna and Marion McLeod have just had their paper on lactoferrin in the skin turned down by *Nature* on the ground that it was an 'odd' experiment."

"Hell," he said. "That's just the kind we should all be doing. But it *is* exciting," he went on. "It holds the possibility of asking many questions and opening up such a neglected area. It is a much-neglected protein, lactoferrin. It *has* to be important. But for Anna it is now a question of refining down—or rather, refining up—the whole process, so that she moves on from correlation and a few experiments to what happens in detail with this disease. Does it do this, or doesn't it? My own hunch is that these things are rather more complex than she imagines."

Two days later Anna received this letter from him:

Many thanks for allowing me to read these fascinating manuscripts. I have no special criticism or suggestions other than those that came up during our nice visit a few days ago.

Frankly I doubt very much that lymphocyte traffic and inflammatory cell margination and emigration are regulated in so simple a manner as you propose here. I make that prediction not on the basis of any special knowledge, or insight, but rather on the basis of experience. . . . In biology it always turns out to be more complicated!

Your hypothesis, even if it turns out to be wrong or overly simplified, is nevertheless an exciting and worthwhile one, for it shows the way to a larger number of well-defined experiments, in vivo and in vitro, experiments that are sure bets to shed light in important dark areas (lymphocyte locomotion and localization, lactoferrin distribution and functions, etc., etc.). This is an ideal circumstance, ensuring a stimulating and productive time in your lab in the weeks and months ahead. How I envy you!

Good luck and thanks again. Please keep me informed of the developments.

On December 11 and 12, 1977, Dr. Henry Kaplan of Stanford, *the* Dr. Kaplan of Hodgkin's disease, visited the institute. He was to spend an hour in Anna's laboratory—Anna had asked Tien-Chun to join them—and was brought there by the head of the lymphoma group.

There was, as is usual on such occasions, a slight air of a State Visit, but they were all looking forward to the meeting. Michael had been particularly pleased with a paper Carolyn had turned up in the literature, in which Dr. Kaplan suggested that the Reed-Sternberg cell—the malignant cell in Hodgkin's disease—was a macrophage. It seemed obvious that he had been seeing the same cells that Anna's group had been seeing, though coming at them via a different route. They all felt very reassured; perhaps, after all, it was not just their imaginations.

"It *is* there," said Anna. "It is not just my impression. As you well know, I explained it to myself—that this is what might happen—that the macrophage is the key. But it is so much nicer to see it in *their* text!"

Therefore, in anticipation of Dr. Kaplan's visit, she had chosen one particular slide out of all their specimens and placed it under the microscope. It was the slide of the cells which in her terms were *the* crucial Hodgkin's disease cells—stained in her way.

Their visitor arrived. From the preliminary exchanges, it was clear that Dr. Kaplan had already been briefed about Anna's ideas. His host and guide said, "You see, Anna thinks it is all iron, and that somehow the spleen traps the cells." And he winked.

Dr. Kaplan was most polite. The whole group was very tranquil and so, too, was the encounter. Anna was clearly determined to play it not only cool but very gently. First of all she asked Dr. Kaplan to "be kind enough to look down the microscope." Then she said, "Are these cells familiar?"

He looked carefully—but it didn't need more than the most cursory examination. "Yes, these are my cells. They are the same ones as mine," he said.

It turned out that he knew in some detail what they had been doing, and he said that his particular criticisms would focus on the care they should take when using commercial samples of antiferritin serum. His group—indeed, several groups—had found that these samples were heavily contaminated with other proteins. Thus they would make the lymphocytes appear to be positive for ferritin when really they were not. The experimenters must therefore make absolutely certain that they had the most highly purified antisera obtainable. He would, he repeated, caution them strenuously, against using commercial antiferritin and against drawing any conclusions from its use.

Dr. Kaplan then turned to a general criticism of Anna's theory. "Does it *have* to be this?" he asked. "Maybe the avidity for iron that you are finding is just one more thing that cells do when they are activated, a secondary reaction and not the primary cause. I think that your study of iron and iron-binding proteins is wonderful, but don't walk too rapidly toward a particular set of molecules. You will find this is a can of worms. The histology [the cellular detail] of the thing is very, very involved. But what you are studying is indeed important, whether or not it is the primary event."

Anna then asked what he would expect to see with regard to the epidemiology—that is, the relationship between the disease and our life-styles and other environmental factors. It was not his field, and he refused to be drawn, although he was aware of studies in Sweden indicating that a twin of a Hodgkin's disease patient could have abnormalities of immunological function similar to those of the patient.

Finally Anna repeated, "Well, I still believe that Hodgkin's disease is possibly a primary metabolic defect, which shows itself either in excess avidity for iron or in the inability of the macrophage to handle the amount of iron thrown at it."

"But why do you start from where you do? Why not start from one cell that turns into a population, transformed in some way into a clone which becomes neoplastic [cancerous]?"

"Even so, you'd still have to ask about the initial event which sets off the changes in the first cell," Anna countered. "I still believe it is the iron situation that supplies the initial abnormality . . . and I take great comfort from the known fact that many metals are carcinogenic."

Dr. Kaplan smiled. "She doesn't give up lightly, does she?" he commented.

Carolyn had a small technical question: What did he find was the best medium for growing the macrophages that she had been trying to culture for nearly two years?

"Agar," he said. "Except that there is a problem. The cells grow luxuriantly and become the cellular equivalent of goats—greedy; they eat too much and die."

The visitors finally left and the tension subsided. The atmosphere changed from one of subdued formality to one of chatty and almost excessive cockiness, perhaps in reaction to their having been so unnaturally quiet and polite. Of course, the reputation of their visitor was so distinguished that politeness was not only the very natural response but the proper one. Furthermore, there wasn't one native American in the room. I doubt whether a group of bright young American scientists would have behaved so quietly.

At any rate, they roared into their post-mortem, all talking at once. "Cause and effect! We were arguing about this a year ago," said Carolyn.

"*Of course* we have considered the impurity of the antisera," said Michael. "We were terrifically careful. We tested all cross-reactions by immunoelectrophoresis. I am, after all, from Bob White's department.* I've been trained to be terribly careful. We always used Scandinavian brands of antisera. Our antisera are very pure. They do not cross-react," he insisted angrily. "We are very thorough."

Carolyn came in again. "It cannot be a clone," she said. "It just cannot. The cytogenetics—the nuclei of the cells—disproves this. The Reed-Sternberg cells are chromosomally not all alike."

Anna turned to me and said emphatically, "That's the construction of the edifice; there you saw people testing our bricks. You must make certain that none are defective. Using impure antiferritin antiserum would have been."

But she was pleased.

"I don't mind about their skepticism. At least they are listening now. The problem is when a glazed look comes over their eyes as it did at the Oxford conference. And look, a year ago we wouldn't

* Department of Immunology, Western Infirmary, Glasgow.

have had anything to say. I'm happy. He spent all his career on this, and now he says it's the changed macrophage. And here we are; we've arrived at the same cell by another route, after *one* year."

One other loose end was being tied up. At the Oxford conference the possibility had been raised that the lymphocytes were dead—killed or damaged by being treated with iron. Furthermore, Anna had been asked how the iron distributed itself over their surfaces: in a pattern that would reveal the presence of an iron receptor or splashed on randomly? These questions had been very much in her mind, but it was not until early January 1978 that a colleague was free to do the first test. The procedure called upon the skill of an electron microscopist-photographer, Pierre, a colleague at the institute and an artist with cells. Photographs taken of cells under the electron microscope always reveal whether or not the cells are alive; when they are dead their ultrastructure disappears. Anna had arranged for Pierre to receive samples of lymphocytes treated with iron—and controls—to prepare, fix, and photograph. She asked him to see if the cells were alive and also to see if there was iron on the lymphocytes' surface. He thought she was quite mad.

One cold winter morning, Carolyn and Nozaki arrived in their laboratory at 5:00 A.M. Pierre had said that 10:00 A.M. was the latest he could receive the samples if he was to process and fix them the same day, and there was a lot of work to be done before the lymphocytes would be ready. Carolyn sleepily rolled up her sleeve and Nozaki drew a sample of blood from her vein. The blood was treated and centrifuged in such a way that the lymphocytes separated out. Nozaki then took the sample of Carolyn's lymphocytes and split it nearly into two; he treated one half with iron citrate and the other with sodium citrate. Placing them in two containers labeled A and B, so that only he knew which was which, he took them over to the laboratory, whereupon Pierre prepared the cells for viewing under the electron microscope: fixed them, mounted them and stained them. Normally all material to be viewed under the electron microscope is stained so that the fine ultrastructure shows up in greater contrast, but Anna and Pierre had decided to omit this procedure in case there were so few iron atoms that the stain would completely obscure them. However, Pierre forgot what they had agreed and went ahead as usual.

Of all the essential tests, this was one of the most crucial and nerve-racking. If the cells *were* found to be dead after the iron treatment, or if the iron was found randomly clumped all over them, then, as Anna said, "We will be totally sunk, and I might as well give up. It would be the end of the theory—and of us."

They waited. At first, they telephoned every week; later, whenever Anna met Pierre, she gave him loving looks, expectant looks, quizzical looks. Finally she didn't know what to try. They lived with this cliff-hanger for four months. And on May 4, 1978, at five thirty in the afternoon, Pierre telephoned Anna. I happened to be in the room when she took the call. With her first words, "Pierre, I don't believe it; how marvelous!" I grabbed a piece of paper and wrote down the half of the conversation I could hear. Anna later reconstructed the other half.

Pierre began by saying, "Anna, I have rusty lymphocytes. Indeed," he went on, "there's ninety-nine percent viability. You gave me two batches, didn't you?"

"Yes," said Anna.

"Well, on one of them there is absolutely nothing."

"Those are the ones with sodium citrate," said Anna.

"The pictures are beautiful. There is iron in well-defined patches on the surface of the lymphocytes, and moreover the stuff is actually eaten by the macrophages. All the cells are healthy, and there is absolutely nothing on the controls."

"Oh, Pierre," said Anna, "this is so crucial. We know there are receptors for iron in bacterial systems. They have been well-defined. But this is the first time it has been shown in mammalian systems. It is terribly important for us. Everybody was so worried about the viability—including ourselves."

"Well," said Pierre, "we are actually seeing the iron with staining, even *through* the stains. It wasn't until after I had done it that I remembered you had said not to stain. I forgot. But you can still see the iron. . . ."

"Oh, my God," said Anna. "That is so nice."

" . . . and I was so excited that I took twenty-four pictures in one hour, and they are all healthy. Is there any difference between B- and T-lymphocytes?"

"We don't know yet," said Anna. "Probably. We already suspect that *both* have receptors. Sebastian is measuring the iron receptors in the mouse. But now we can proceed to separate T- and B-cells and work out just what the receptor is. Some viruses utilize the iron receptor in bacteria in order to get into them. Now, I tell you, it is going to be the same in mammals! And when we treat the cells with iron citrate, we get inhibition of mixed lymphocyte reaction. This may be a very important receptor indeed."

"What do you want to do next?"

"I think we *have* to go to the X-ray diffraction microscope to show that the metal *really* is there. But you should do it," she went on. "You are the first. Let us try to get the electron-probe analysis done soon, to prove that it really is iron. And this receptor on the lymphocyte is going to be the receptor for the virus too, I am absolutely sure."

She hung up and then called Nozaki. The phrases rushed out. "I have a present for you. . . . Pierre found patches of iron on the surface of your sample. . . . You must go there and talk to him. Moreover, the macrophages have eaten the iron. . . . He found that it had been ingested by the macrophages. Go now and talk to him . . . just go there."

This was one instruction from Charlie's Angels that Nozaki was very happy to receive. Anna was rushing on. "Oh so many congratulations. He is so excited. We can do things. We can separate out the Bs and Ts. Do you remember we said not to stain? Well, he forgot and in spite of that he could still see the iron. Go now, Nozaki."

She put down the phone and turned to me. "It never ends," she said blissfully. "You see, at last we can really test my Hodgkin's disease ideas. There are lymphocytes with iron on the surface. There are monocytes—the macrophages of the blood—that eat iron, and that is bad. Polymorphs have lactoferrin, so they can take up excess iron, all right, and so immobilize it. So if you have lactoferrin you're all right. But if you don't, the iron can get into the macrophage, and it eventually may become a Reed-Sternberg cell."

"But," I said, "these receptors you keep talking about. How does the iron attach to them? Is the receptor a thing like a little basket?"

"Oh," said Anna, laughing. "A receptor is not a *thing*. It is an embrace. It is a protein that embraces. . . . I think it embraces many things. It embraces lectins; it embraces viruses, many things. And if it really is produced by the HLA locus, as I think it is, well . . .

"You see, Pierre will be able to help us. We can give iron to the system and Pierre will be able to see just where the iron receptor is on the surface of the lymphocyte—whether indeed it *is* the HLA receptor. And I tell you: It either *is*, or it is right alongside—I know it. . . . And if either is true, we'll have champagne."

The HLA locus—the histocompatibility locus—is *the* most important complex in immunology, so important that champagne was undoubtedly the only drink in which to toast it. Its function is analogous to that of blood groups. Everyone's blood can be labeled Group A, or B, or AB, or O. This means merely that our blood produces certain proteins, and you can only receive blood from someone in your own group, whose blood produces the same proteins. Given blood with foreign proteins you will die, because they will coagulate your blood—which was seen very quickly in the earliest attempts at blood transfusions.

It is exactly the same with lymphocytes. They, too, produce surface proteins, roughly four or five specific ones, and so human beings are grouped accordingly. You can accept a kidney or bone-marrow transplant from someone—usually a relative—whose lymphocytes produce the same proteins. But you'll reject such a transplant from others, because your lymphocytes will recognize their proteins as foreign and will attack the transplant. A test called the "tissue-typing test" is always done before a transplant. It "types" the proteins of both individuals so as to match up donor and recipient as closely as possible. They must be compatible; hence the phrase histocompatibility—HLA, for short.

The crucial proteins are produced at the HLA receptor on the surface of the lymphocyte, and their production is controlled by a group of genes inside the nucleus of the cell. And now Anna was saying that the receptor for iron on the lymphocyte surface was likely to be identical with the HLA receptor, or right alongside.

"You're hedging your bets, aren't you?" I asked. "Saying that they might be next to each other."

"Of course," said Anna. "But of course. You should always have two alternatives in science so you are never depressed if one of them doesn't work! But you know in this case we had *no* alternative. If the cells were dead, that really would have been the end."

"Do you suppose there are receptors on lymphocytes for *all* major metal systems?"

"That's not a bad thought, merely an antiquated one. Yes, of course! This whole iron story is fantastic. Just think of the virus for encephalomyelitis: Its active part *is* iron. You can replace the active part of the virus with iron and still get the same effect. And now I tell you that malignant tumor cells can eat iron. I tell you they are going to take it up like water. *All* of them. They positively drink it."

CHAPTER 12

SOLITARY TREES

Creativity is what cannot wait, cannot stop, cannot backstep: faster or slower, it always goes ahead—through, alongside, above, regardless of crises or systems.

JOSÉ RODRIGUES MIGUEIS, Address to the Center for Portuguese and Brazilian Studies, Brown University, March 11, 1979

By December 1977 Anna felt an acute need for reassurance, for some outside corroboration, and Pierre's confirmation of her predictions was still months away.

"This," she said, "is the most miserable time of all. The hunch is stronger than the observation. The links between hunch and observation are clearly there, but at this stage people still think you are quite mad. They are as fearful as if you had cancer. The danger in cancer is to die; the failure in this madness is the failure to communicate. And the danger in the scientific process—at least in this society—is the danger of having an idea which is wrong. It's equivalent to death."

"Scientifically approved but not funded," continued to be the recurrent label on her grants. "You know," she went on, "I'm really going to rewrite Descartes. He said, 'I think, therefore, I am.' Here you must say, 'I produce, therefore I exist—but only if I produce.' Do you know what I would like most of all as a Christmas present? The knowledge that someone, somewhere, in the world—just one person would do—is thinking along the same lines. Just one person.

"One is totally alone. It is like the thymus-dependent areas all over again. Why should Dr. Vera have believed me against the whole world? She only believed me when the photographs were published in that other article, and there the areas all were, but the other people were not seeing them. The fact *was* there; then I really *saw* it; but everyone else was cautious, cautious, cautious. And now, once again, the whole picture is clear in my mind. I see the whole thing with the clarity of Caribbean water; I think everyone else is looking into the muddy waters of the East River, and this is a dangerous madness."

She continued to feel that her ideas were generally right and that other people would eventually get to the same point. Some might, indeed, already be there. But there was no sign of any "Christmas present," no signal that somewhere in the world someone was thinking, if not actually saying, the same things.

In late December and early January, everyone—except, of course, Carolyn—had flu at least once. Nevertheless, by late January Anna had set in train two separate scientific ventures which started off very promisingly. Each had fanned out from a facet of her main inquiry, and the outcome of each would eventually prove vital to the success or downfall of her ideas. I was away—four relentless weeks of flu had forced me to go off to recuperate—but during my absence we kept in touch through letters and telephone calls.

January 7, 1978

Perhaps God, at least the section of heaven that looks after writers of science, never intends you to be around when significant things happen. For perhaps the most significant thing since we started in this venture happened yesterday: One of Tien-Chun's patients, J.F., aged nineteen, died.

On the same day Tien-Chun had an almighty row with a colleague who, at the end of three years of never looking at the data, finally did look, saw its obvious interest, and became very possessive. To ask Tien-Chun about her day yesterday would be a marvelous document of the contemporary triangle in cancer research: the dying patient and her family; the impotent clinician; the greedy researcher in the laboratory, wanting the data for his (her, women are no different) abstract, paper, or grant, or anything to do with himself or herself.

Otherwise I did lots of happy things today. Went to the hairdresser, bought a Neruda book, had lunch with Tien-Chun and Ruth, looked at things down a microscope. Actually this book of Neruda's is his saddest. Published after his death, it portrays a man at the end of a journey, asking questions about death and the autumn:

> *Do you know where death comes from?*
> *Does it come from above?*
> *Does it come from below?*
> *Does it come from microbes or from the walls?*
> *From the winter or from the wars?*

Today, I say, on the death after nineteen years of J.F., the painful answer, the painful answer is from the walls . . . from the walls of our ignorance and the persistence of our arrogance.

On February 11, Anna telephoned to say that Father Christmas had finally arrived. She had been very busy with meetings, experiments, grant applications, God-knows-what, and also there had been a seminar to attend, given by Dr. Hoffer of Yale. She hadn't really felt like going but thought she should. She went and for forty minutes regretted it, because his lecture was very empirical, formal, at times even a little dull. But it would have been rude to slip out, so she stayed. Dr. Hoffer was talking about gallium, and, in nearing the end of the hour, he said, "Gallium-67 may be useful to trace those *tumors which take up iron.*" Anna, stunned at the matter-of-fact mention of tumors taking up iron, instinctively put out her hand and found it was being clutched by one of the clinicians whom earlier she had battled fiercely to convince. On February 13th, she wrote me about the occasion:

Well, as I promised, here I am watching a television program about Martin Luther King. But I will try to tell you about Dr. Hoffer's talk in the in-between bits (otherwise known as commercials). In summary, Dr. Hoffer got to where we are but along the gallium road. You see, gallium is in many respects similar to Fe [iron]. So he built up his talk to the point where he actually said, "Gallium may be useful in tumors which take up iron." Nothing special about that; bacteria do. But, you see, among "the tumors that take up" gallium is Hodgkin's disease (!), as well as lung carcinoma, testicular carcinoma, and some melanomas. And he went on and on, about things that you by now are thoroughly familiar with: receptors for iron in bacteria, viruses using the iron receptor, etc., but none of my clinical colleagues knew about any of this. He even talked a little bit more about lactoferrin than ferritin. It is so nice.

I'm afraid I don't feel like saying much now. Above all I want to go to sleep and cannot. Go to sleep for a week, two weeks, three weeks, and then come back sure.

February 14, 1978

Last night I had the fatigue one has when finally one can go on holiday, and I did. I slept very well and woke up hardly believing that it has happened. You may also understand how between the fatigue and Martin Luther King, and Michael with his financial problems, iron and Hodgkin's disease did not seem so overwhelmingly important. And with the weight of iron advertisements in this country one inevitably starts feeling that it is not going to be easy to tell this community that in some circumstances iron may be bad.

It is iron and flint stones, and the food almost all of our Hodgkin's children ate—as we found out from the rough survey of eating, drinking, and living habits. The blacks finally got their civil rights—their voting rights—and, with all those rights, the rights to sell things on T.V. You remind me often that it was in this country that I made these discoveries, all these pressures notwithstanding. O.K., I did. But part of my extreme tiredness is the effort not to be prostituted by the system. Not to sell myself, my beliefs, my standing, my honesty to get the grants I must for these kids who have been working with me. I will not say in a grant application that I can anticipate a result when I cannot.

You tell me there is hope. I will tell you the biological lessons of hope I am starting to think about. How the concept of death really is just the nostalgia of the beholder. How in evolution the basic units of life were never eliminated [proteins and amino acids] but were incorporated into higher organisms. How during the control of the differentiation of blood stem cells, the little polymorphs tell the macrophages what to do, who in turn tell the stem cells what to do.

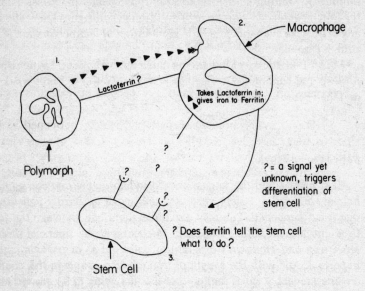

A signal, as yet unknown, triggers the differentiation of the stem cells in the bone marrow. But I think Edward and Michael have cleared up step one, the message between the polymorphs and the macrophages. It is exciting, because once you know steps one, two, and three, you really will be able to treat leukemia gently.

The ballet of molecules and membranes, if we could visualize it, must be one of the most beautiful things to be seen. The biological world is a gentle one. Cells do things with the best of intentions. To kill the tumor cell with the killer instinct will not work. It sounds good; it makes good fund-raising speeches. But it is not right. Dr. Hoffer knew that. He mentioned that some people had worked at Harvard with antibiotics in 1940 who did not go further because the antibiotics did not kill the bacteria. The National Cancer Institute is doing the same with tumor cells, he said. It wants to test substances that kill, not substances that stop growth. And it may be wrong, the wrong tactic again. It is clear that Dr. Hoffer is already playing with iron chelators in mice. I cannot wait until the first of March, when I have a date to go and have a long talk with the deputy director of the institute.

Early in March 1978, almost three years to the day from the start of this project, Anna and I embarked on a summing up. It was, she said, the end of a time. The main thinking had been done and, as she wrote later, "The high tide of 1977, with its scientific experiments and its personal loneliness, has run swiftly back into the sea." From now on it would be a matter of picking up treasures on the shore.

"Nothing in the immediate past," she said, "happened as expected, and I have learned a lot on the way. I have now an idea of the circulation of the lymphocytes which may be entirely wrong. It is certainly incredibly oversimplified. But more important, really, is my theory that in essence the immunological system has evolved with the ability to survey, and recognize, and utilize—I don't know exactly what. But this ability is expressed as a capacity to survey metals and, in particular, iron.

"My reasoning goes in a simple way. After aluminum, iron is the most common, simple metal and constitutes one third of the earth. It is also present in up to forty percent of meteorites and so was one of the very early elements in cosmic evolution. Its geochemistry is fascinating. But what is even more fascinating is that bacteria have developed extremely efficient systems to use it. This has been most beautifully worked out in the microbiological field. But in the higher cells it has not. Yet I believe the same systems apply equally, and this is how I think it works.

"So far as iron metabolism is concerned, much is already known. We have a mouth, a stomach and an intestine, and normally most of the iron gets into our system this way. It goes in as ferrous iron, and is then rapidly oxidized into ferric iron. This is then bound to a protein, transferrin, and so is carried into the bone marrow and thus given to the early red blood cells, which need it for the blood cycles. So the red cells have their own pool of iron.

"But iron is also bound to a second protein, ferritin, which is a storage protein, and so we also find the metal in "storing organs" like the liver and the spleen. When red cells get old they break down in the spleen and are eaten up by the macrophages, which then make ferritin. In this way the iron is turned back into ferric iron and enters the storage pool again.

"Then there is the third protein, lactoferrin, which is synthesized by polymorphs, the white cells, and is present in milk and in our secretions. It has three hundred times more avidity for iron than transferrin. The whole iron system is an extremely clever cycle, because very little iron is excreted. The amount we excrete per day is absolutely minimal. You can only shove it in; you can't push it out. And so an organism has control of iron *only* at its point of entry, for there is no excretion mechanism.

"Now if, for some reason or other, there is too much iron around, the macrophages go and mop it up. And as far as I'm concerned, in the diseases we've been studying—Hodgkin's, leukemias—there is an abnormality in the lymph nodes and the macrophages in regard to the intake, or the handling, of the iron. I think that because the lymphocytes have iron receptors, they too can get captured in the spleen by the two iron-binding proteins, lactoferrin and transferrin.

"All has an evolutionary background and relates to other biological systems in this way: As you know by now, bacteria can get infected by viruses, which enter the bacterium and tell the bacterial DNA what to do. But before this can happen—before the virus can get into the bacterium—it must find a surface receptor to latch onto, and *for this purpose it uses the iron receptor on the bacterial membrane*. People have found that certain varieties of the bacterium *E. coli* are resistant to a certain virus. Well, the reason is that these varieties don't *have* iron receptors, and the virus, having nothing to latch on to, can't get in. Equally, if you have masses of iron around, the iron sits on the receptors and blankets them—and again the viruses can't get in. So by using iron you can actually stop the virus from infecting the bacteria.

"So we already know that infecting viruses get into bacteria by utilizing the iron receptor . . . and we also know that bacteria need iron for their respiration in order to become infectious and grow in our cells. What I am now saying is that all this applies to the higher vertebrates too. It seems to me very likely that the immune system must have evolved in ways that make it possible to combat this very clever, very nasty bacterial system.

"My thought is that maybe the traffic of lymphocytes through the body, and the migration of polymorphs, and the functioning of macrophages are *all* related to this system and are all controlled in part by the presence of iron. I don't know. But I do insist that in higher organisms the immunological system evolved to cope with the problem of bacterial infection, along with other problems—to cope, that is, with a system that is iron-dependent or iron-related.

"Take lactoferrin, with its tremendous avidity for iron. You can see how it could control infections. At least I can . . . easily. It is produced by polymorphs, and because it drinks up iron, it takes the metal away from bacteria. But lactoferrin can also bind to the macrophage, so that the circle of control goes like this: The polymorph releases lactoferrin; the macrophages have receptors for it, they take it up and carry it around the body; wherever excess iron is found the lactoferrin mops up the metal, which is then not available to bacteria, and so they can't become infectious.

"So I think we have a system of 'proteins-with-metals' which has evolved for very precise functions. The immunological system is extremely economical: It can function with just the three white cells, the three iron-binding proteins, and virtually one metal—perhaps two; polymorphs can synthesize lactoferrin; macrophages can synthesize ferritin; lymphocytes have iron receptors. These three types of white cells are geared to detect the presence of metals—and not only detect them but utilize them, and probably utilize them in the control of very fundamental processes. With just these you have a tremendous control, and this control is *absolutely fundamental*.

"Thus, as a result of this very messy year—which in respect of thinking was extremely messy—we are down to extreme simplicity, and this is exactly where I like to be.

"As far as lymphocyte migration goes, I have a simple scheme, again probably too simple. Lymphocytes go and are caught where there is an excess of iron. But in addition, I now have this other question: Do the lymphocytes actually migrate because they have a receptor for iron or one for the protein? Just what is their receptor for iron? By now, as you know, I *think* it is—or is next to—the HLA, the most important group of surface antigens so far as immunologists are concerned. The proteins it produces determine our tissue type, which determines whether our organs match up with other people's. But the HLA complex couldn't have evolved simply for the sake of receiving heart transplants. So why do we have it? I say we have it, at least in part, to cope with infection and to do so *indirectly* by holding onto iron.

"One last thing: It is known that some virulent strains of bacteria have become so virulent just because they have acquired a tremendous avidity for iron and suck it up much more than other strains do. I'm saying tumor cells are similar. A malignant cell, like a virulent cell, behaves like a virulent bacterium. It becomes capable of sucking up iron avidly and utilizing it. The tumor cell then starts dividing like mad—as bacteria do. I've said that I think the signal for lymphocyte circulation is the presence of iron. But if the tumor cells eat up the signal—again it's iron—the lymphocytes won't go there. The tumor will grow undisturbed because the lymphocytes have not had a signal to move toward it.

"Iron is incredible. A recent paper in *Nature* described activated macrophages that have the capacity to kill human tumor cells, but if the macrophages are offered red cells, they gobble them up instead and then cannot kill tumor cells. The scientists broke down all the components to see what it was in those red cells that was causing the trouble, and they came to—iron! They found that if a macrophage is loaded with iron, it *cannot kill a tumor cell*. And one of the things I'm saying is that the process of malignant transformation *includes* becoming highly avid for iron. In other words, if you have too much iron in your system you are really laying yourself open to trouble because, on the one hand, the tumor cell will utilize the iron for itself and, on the other hand, since iron is also present elsewhere in the body, the lymphocytes may be distracted. The macrophages, too, will be loaded with iron, and therefore *nothing* will be available to kill your tumor."

"What, then, at the end of all these speculations are your actual predictions?" I asked.

"I have four," she answered. "First, these iron-binding proteins are going to play a role in lymphocyte traffic. Second, they are going to be important in the functioning of the immunological system, in relation to the activity of viruses and bacteria. Third, they are going to be important in controlling the red blood cell system because, crazily, I think lymphocytes probably exchange electrons much as red blood cells exchange oxygen. Red blood cells make hemoglobin. To do that they need iron. Lymphocytes latch onto iron. Somehow, I feel it's all connected. And fourth, I think all this is going to be very important for cancer. I think tumors need iron in order to help them survive and that malignant cells acquire a receptor for it.

"I also predict that 1978 is going to be quite different from

1977! Even though no one believes me, much of this is no longer an idea. It is a *fact*. You know, scientific activity is just like any other activity, and ninety percent of the time everything is lost—not kept, not recorded, not remembered. It is lucky for me that you've been around because now I remember only the clever things. I don't remember the way we floundered around last year."

"The blunders wash off, don't they? In the rivers of your days?"

"Yes, indeed. It all seems so very clever now. If I were to reconstruct it I would reconstruct it totally inaccurately. I certainly could not have predicted where we are, and I cannot remember how things went, nor really how we come to be doing what we are doing."

CHAPTER 13

DIVIDENDS FROM
TIME'S TOMORROW

Everyone suddenly burst out singing...
SIEGFRIED SASSOON, "Everyone Sang"

What they were doing in March 1978 was pursuing the two simultaneous projects that had been set in motion in December and January.

Michael was collaborating on one with Edward, a graduate of New York University who had received his doctoral training in cell biology and whose thesis had been devoted to hematology, the study of blood, which fascinated him. Initially he received very little encouragement to continue in the field, the general advice being "You are not an M.D. Get out of hematological research; it is the domain of doctors." He ignored the advice.

He began his research career in the department of medicine at a university in Canada, studying the "committed" stem cells in human bone marrow. The young stem cells in the bone marrow are the most vital for our blood and immune systems, for they are the "beginner cells" from which all the others descend, whether red blood cells (erythrocytes) or white blood cells (polymorphs, macrophages, lymphocytes). Deep in the core of our bone marrow these stem cells divide to form the progenitor, or "committed," stem cells, each of which will be restricted to the formation of one or possibly two types of cells. From them all the cells of the blood and immune system will eventually grow and mature.

But there clearly must be a limit to such growth. Indeed, uncontrollable bursts of growth resulting in abnormal numbers of white blood corpuscles are the hallmarks of leukemia. How then is the process of blood cell proliferation started, and how is it stopped? How do the committed stem cells get switched off, and what is the nature of the signal that switches them off, chemical or physical?

Luckily these committed stem cells can be studied in culture dishes, where they will ultimately grow into colonies of specific blood cells.

Over the years Edward had found that in culture this colony formation is at least partly controlled by a simple feedback mechanism and that chemical substances are the controlling factors. While polymorphs (granulocytes) are developing they begin to synthesize an inhibitory factor, which is stored in the mature cell. This inhibitory factor does not act directly on the stem cells of the marrow. The monocyte, or its descendent cell, the macrophage, acts as the intermediary. The mechanism works like this: Polymorphs produce substance A, which acts on the macrophages; they then respond by *ceasing* to produce substance B, which normally triggers the stem cells to produce new cells. When this trigger stops, owing to the cutting off of substance B, the "committed" stem cells stop producing more colonies of polymorphs. But in chronic or acute myeloid leukemias, the first substance, A, is either missing or present only in very low concentration. Thus nothing stops the macrophages from manufacturing substance B, and therefore the stem cell continues to produce bursts of polymorph colonies. Because indirectly substance A acts to inhibit the production of these polymorph colonies, Edward called it the Colony Inhibitory Activity factor, or CIA for short.

In 1975 Edward came to the institute as an Associate Researcher and joined the laboratory that specializes in problems of blood development and regulation. There, as he says, he was "lucky to hook onto" its head, "one of the best minds in the business and a damned nice man." Soon, Edward had shown that he could get the same inhibitory effect on the stem cells without using polymorphs at all. An extract from them, or even a culture medium in which they had been immersed for two days, would do just as well. But the problem was: What *was* this CIA substance that acted on the macrophages? In collaboration with the biochemistry section of the lab, Edward was slowly trying to analyze it, purifying it stage by stage, studying its role in animals, and analyzing in detail the effects of its absence in leukemias. But he was getting nowhere with the basic question: What was it?

"We were like detectives looking for a murderer," he said. "But we had no real description to go on. We were following up every clue, biochemical or physiological; we were purifying the CIA factor, stage by stage; but all we really knew about it was that it had one known function. Now, you can say to your colleagues, 'I've got a molecule and this is what it does,' and they will go along with you for a while. But eventually, and within a decent amount of time, you should be able to show exactly what your molecule is."

He went on to say, as others have said, that success is compounded by hard work, luck, and timing, whom you happen to meet and when. You can, when you are young, try out your ideas on an established person only to find that you have come up squarely against a malignant personality who will say that your work hasn't proved anything, you've not done what you thought you'd done. This can be devastating to a young scientist, and when one looks up from below at someone who presumably "knows it all," one's reaction is to sulk, or give up, or just show them.

"On the other hand," Edward continued, "there are very good people who have made significant findings themselves and who encourage you thoroughly, who explain nicely to you why or where or how your work is deficient, who give you constructive criticism and direction and conditions for going on. But you have to be a little tough. You dare not give up, because if you do you lose all. If you hang on, you might be wrong, sure. But on the other hand you might be right, and you have to go on with what you believe.

"That's why it was so nice when Anna brought you into the lab a year and a half ago and said in front of you what she said about my work. That was one of the nicest things that has happened to me in my time here."

Anna had liked his work from the very first moment she heard it described at the site visit she attended in May 1976. She believed his discovery—that in the blood cells of leukemia patients the Colony Inhibitory Factor is present only in very low concentrations—to be absolutely fundamental. After the site visit ended, she had introduced herself and told him how elegant his whole concept of regulation appeared to be, and a few days later she telephoned to repeat that she considered his work among the very best being done at the institute. For eighteen months after that, both went their own ways, Anna along those paths that led her to iron and the iron-binding proteins, while Edward continued trying to work out what his CIA substance was and whether it functioned in live animals as it did in laboratory cultures. But Anna never forgot his work, and one day, when she was ready, she asked him to talk to her people.

A weekly Tuesday meeting, very small and very informal, is a regular feature in Anna's group. Members and colleagues slip in for a couple of hours to discuss their work. On December 13, 1977, Edward gave them a brief talk. When he had finished, Anna, who had listened intently, asked if he knew anything about lactoferrin. He replied that he knew nothing at all about it but supposed it had something to do with milk. Whereupon Anna said, "Well, I think lactoferrin is your CIA factor."

"My immediate, private reaction," he told me, "was 'Oh, bull-shit!' But I just said, very politely, 'That's very interesting. What makes you think so?' Then she began to tell me. She obviously had a clear picture in her mind of the feedback triangle of regulation I had outlined—the one that controls the growth of the cells produced by the committed stem cells—and she clearly understood the concepts and the biochemistry of all the molecules that were involved. But the whole group started barraging me with facts and with papers, and the more I read their papers the more I really didn't believe it. For it was totally unbelievable, just too good to be true."

Edward's skepticism had two sources: the reflex skepticism of all scientists, and the fact that it really *was* too good to be true. One never expects solutions in science to appear out of the blue, or to be handed one on a platter, and he had been resigned to many more years of work before he might possibly find out just what the CIA factor was. In addition, the connection appeared to have been made far too easily. It was as if Anna had reached up into the air and pulled down—lactoferrin. The link between the two people and the two ideas was, others were to say later, "fortuitous."

The connection had not, however, come to Anna at all easily and had certainly not been plucked out of the air. The only fortuitous element was that both scientists were in the same institute at the same time. Anna had been captured by the intellectual possibilities inherent in Edward's problem the moment she heard him speak about it; she had already speculated that her own lines of work might be relevant to leukemia; she had been thinking ever since May 1976 about this feedback loop of cell regulation and, later, about its possible relation to the iron-binding proteins. She had spoken to Dr. Hirsch and to other colleagues whose past work bore upon her present problems; she had ranged widely through the literature, pulling out what she considered relevant and interesting. In one place she had found reports which showed that polymorphs synthesize lactoferrin at sites of inflammation. In another she learned not only that macrophages have receptors for lactoferrin and follow the polymorphs to inflammatory sites but that their accumulation at these sites somehow depends on the release of lactoferrin by the polymorphs. So if Edward's question was reworded: What is it that polymorphs produce that might be taken up by macrophages? there was at least one possible candidate—lactoferrin.

At the Oxford conference three months earlier, Anna had speculated—albeit somewhat tentatively—about the relationship between polymorphs, monocytes, macrophages, and lactoferrin,* raising questions that had been in her mind for some six months, worrying over the odd biological coincidence whereby all three types of white cells in the circulation participate both in the recognition and in the synthesis of the three iron-binding proteins: lactoferrin, transferrin, and ferritin. Gradually, as the days passed, she had put the pieces together. It was then—when she was ready—that she had invited Edward to give a talk to her laboratory. Even though his problem was not hers, nor his questions her main questions, her ideas were coalescing with his.

Edward may have been skeptical, but he did agree to do *one* experiment. He insisted only that if it didn't work, that would be the end of the matter and they would get on with their respective jobs. Shortly afterward, samples were exchanged. Edward gave Michael his CIA factor and in return was given lactoferrin and the antiserum to lactoferrin from Michael's supply. Then they all went back to their own laboratories and repeated their own range of experiments with the other's substance.

At that time Edward's wife was pregnant; so too was Michael's. This perhaps was the truly fortuitous event in the whole episode, for after the children were born the mothers were able to assure an adequate supply of human milk, and thus of lactoferrin, for continuing the work. Two weeks later, Edward came into the lab saying just what Michael had said a few days earlier: "I've had a son, and here are the results."

---

* In her paper published in the conference proceedings there is a diagram showing this.

If, at first, Edward really couldn't believe that his CIA factor and lactoferrin really were the same, after the first batch of tests he believed that they *might* be. From then on both scientists moved very fast. Experimentally they did everything they could as cautiously, as carefully, but as completely and quickly as possible. Within three or four months they had almost definite proof that the two substances were indeed identical.

To discover whether the two substances were the same, they had attacked the problem in three ways: They did experiments which showed that lactoferrin and the CIA factor acted in the same way, even at such phenomenally low concentrations as one part in a billion; then, bearing in mind the possibility of contamination in a complicated experiment, and having ascertained that their antiserum was specific for lactoferrin and absolutely nothing else, they went on to show that throughout the whole range of their separate and joint experiments, this antiserum would smother the activity of both substances; last, using another reliable biochemical procedure, they showed that both substances were active in an identical range of acidity and alkalinity.

Soon they were doing the final series of double-blind studies, in which each experimenter was given coded, unlabeled samples. As they went through the total range of experiments once more, neither Edward nor Michael knew which substance he was testing.

"I was really scared to do that lot," said Edward, "even though, actually, we had done many of the earlier tests double-blind because so much was at stake. And I was so scared that it wasn't going to work that, when I brought the figures down to Anna but before I knew the results, I said, 'If this is not lactoferrin then I'm stopping right now.' And there was one really hairy moment—a point when I was sure we'd blown it. Michael gave me two different forms of lactoferrin—one saturated with iron and the other unsaturated—and told me he thought the unsaturated form would work better, have the stronger inhibitory effect. When I did the experiment, I saw the opposite. It was the saturated form that was more effective. The data were so good that I couldn't believe we were wrong, but it was the exact opposite of what we thought we should have seen. So I went downstairs and I said, 'Well, that's the end. We blew it.' We sat about for an hour, just not understanding it. Then Nozaki came by and we rehashed it all for him, and he said, 'No, you've got it all wrong.' And he went on to explain that up to this point the only role postulated for lactoferrin was inhibiting the growth of bacteria, and it is lactoferrin *un*saturated with iron that does this, immobilizing bacteria by taking up the iron they need. But the form of lactoferrin which binds to macrophages is the iron-saturated form.

"That tied it all together. At that point I was thoroughly convinced we were on the right track. Not only would we be able to put out a paper saying, 'This is the CIA factor we've been talking about these past three years,' but we would be beginning to understand how it works, and how its capacity to function depends on its binding to iron. And that was really good because there already were relevant papers from a Belgian group who had done such a fabulous job, and who had shown that lactoferrin only binds to monocytes and macrophages *when* it has iron. So we were able to put everything together and show how all the pieces fit into the puzzle."

Anna says of that triumphant effort, "This shows you a place like the institute working at its best, a place where, if you will go and find them, there are all the people you need doing relevant and valuable things, with whom you can collaborate and pull things out quickly, if you all choose."

Edward says, "If Anna had not seen the connection, it could have taken me anywhere from a few years to never. She was the key. She was the person who remembered my work and then at the right moment pulled out what was needed. It is unlikely that I was going to make this connection because I was simply not looking that way. However, at a point at which Michael and I had quite a bit of data, the head of my laboratory was discussing my work with other people at a conference. And someone said casually, 'Has Edward ever considered that his inhibitor might be lactoferrin?' And my boss replied, 'He is trying a number of things.' I suppose that if at the time I *hadn't* tried lactoferrin, he might have told me to do so. But then, on the other hand, he mightn't have. So who knows what would have happened?" Finally, Edward said simply, "Of course, I hope the concept holds up and I hope other people will try to duplicate the work."

"What about therapy for leukemia?" I asked.

"We don't know *anything* like enough yet," he said. "Maybe it can only be a test for remission. Maybe when a leukemic patient is getting better it's because the cells are producing the lactoferrin and thus controlling the uninhibited multiplication of the other cells.

We don't know. But at last we now have a defined molecule; we know what it does, and this opens up unbelievable areas of work with clear directions, and these directions obviously lead directly into leukemia. Six months before, I was working with crude material somehow trying to clear this molecule up. Now we know what it is and have it in purified form. One of the things we obviously must do is to check the points where, in leukemic patients, the feedback triangle that should control production of these cells breaks down.

"But still it is just unbelievable. Because it works in vivo—in animal experiments—and at such low concentrations, we really believe that lactoferrin is a physiological regulator that controls the production of polymorphs by acting on the macrophages."

Their paper, with the fullest experimental detail, was finally published in the *Journal of Experimental Medicine*, eleven months after Edward had given his talk to Anna's group. Two months after the paper was published I spoke with Edward again. By then he had given papers and reports on the work to three separate conferences. In the interval Michael had resigned and returned to Britain, but Edward had continued along the lines they had opened up and had begun to sharpen the details of their studies, confirming, to begin with, that the function of lactoferrin was totally specific. The problem with biological inhibitors is that people tend to think they can inhibit just about everything, and so Edward had tried lactoferrin on all known functions of the macrophage to see if any others were affected. With one exception they were not. There were still, of course, open questions about the precise mechanism whereby lactoferrin plays its role and about the nature of the receptor on the macrophage to which it binds.

But the range of Edward's and Michael's experiments had proved to be reproducible. The series had been duplicated independently by another colleague, and this fact, together with their original data, the confidence that Edward's boss had in the work, and the general sense it all made led to a not unfavorable initial reception. Even so, Edward said, "It is too early to be sure of its acceptance. I learned a long time ago that most people never really tell you what they think. If they think it is a good paper they don't say so, because they assume you know it. And if it is not a good paper they say nothing, because they are not interested enough and they think, Hell, why tell him anything?

"The only way we are going to know what people think is when they are interested enough to do the experiments themselves. And they've not been in the literature long enough for that. Other people are either going to see the same things we see, or not; either believe us, or not. I'm very confident that if people will do it the way we said it was done they will see the same things.

"People have come up to me and said they believe it—they really believe it. But there are problems. They find it hard to believe that anything can work at such low concentrations. It's unbelievably low. They are satisfied with the concept; they are satisfied with the data. But they cannot understand how it can work at such low concentrations.

"So the first criticism came here: Was this going to be another case like the nerve growth factor, where at first it was reported that there was an effect at very low concentrations, until it turned out that the technician doing the experiment was dragging the pipette over from one solution to another and taking some of the original solution along? In other words, what they thought was a dilution wasn't a dilution at all. That was the very first thing my boss asked me. Well, that's not our problem. We changed the pipettes when we did the dilutions. So if there *is* something wrong about the fact that lactoferrin works at such low concentrations, I don't know what it is that's wrong.

"The other problem is this: If you assume that each molecule of lactoferrin hits randomly, then, in order for the effect to be produced at such low concentrations, it must mean that only one molecule, or even less, hits the macrophage to produce the effect. And people saw this as utterly ridiculous. One person commented publicly that at this dilution it would take about three years for the molecule to hit its target macrophage. This could have been nasty sarcasm, but I took the comment and fielded it because I agree. It was unrealistic, based on everything else we knew at that time. *But it didn't mean we are wrong*.

"People say that cholera toxin works at very low concentrations, but I don't know how low their 'low' is. And Michael told me about newly discovered substances one molecule of which, at a distance of one mile, can turn a whole salmon around, as it migrates toward its parental river. But there they are not talking about mice or men.

"The thing is that the critics are assuming *random* hits, and we don't yet know if that's the case. Maybe it is; maybe it isn't. Maybe it is not random; maybe it is actually a case of directed motion of the lactoferrin molecule on just certain specific macrophages. But right now there is not much evidence one way or another. We were assuming initially that every macrophage in the culture dish *is* a target for lactoferrin, but that, too, is far from the case. We—the people in the lab, Anna, and I—now have quite a lot of repeatable data indicating that this is not the case. We are separating out the macrophages that we use as 'target cells' and trying to see, first, whether lactoferrin hooks up to *every* one of those cells or only to a certain fraction of them and, second, whether lactoferrin inhibits in *every one of these cellular fractions*.

"We already know that lactoferrin does *not* hook up to every macrophage, and we know that only a very small percentage of macrophages are target cells for lactoferrin. So in actuality you are probably getting more than one molecule of lactoferrin hitting the critical cell. Perhaps in the range of ten or twenty or a hundred are available for each target cell. This puts you into the range of probability for the event. Or maybe, as the head of my laboratory suggested, you just need the first hit and then it is a cascading system. One molecule hits one macrophage, which changes everything else, like knocking over dominoes. We don't know yet."

"Where are you going from here?" I asked.

"We want to know *exactly* how lactoferrin works. We are really only at the infancy of understanding what is happening. It is just the beginning. For example, are there really receptors specific for lactoferrin on the macrophages? The work by the Belgian group suggests that there really are.

"The second problem is, what exactly is lactoferrin doing once it binds to the macrophage? The only thing that's known is again based on the work of the Belgian group. They showed that it releases iron for the other iron-binding protein, ferritin. We also know it inhibits the production of the factor the macrophages produce, which in turn acts on the stem cells. But scientists are only at the very beginning of understanding all this, and we have to wait until this area becomes more defined."

CHAPTER 14

POSSIBLE WORLDS

Nature is no spendthrift but takes
the shortest way to her ends.
EMERSON, "Fate"

In the autumn of 1977 Anna had said, "I believe lymphocytes are
deeply involved in the iron cycles. But how on earth am I going to
prove it?" By then she was steeped in the problems of iron and iron-
binding proteins, bacteria and bacterial infection, and the evolutionary role of the immune system in relation to both. Having asked
a whole series of questions about iron and the immune system, and
having linked them together in the most fundamental biological
context—that is, their joint development through historical time—
she had thus provided herself with both a theoretical framework for
"making sense" of the phenomena and an agenda for future experiments. Her letter to Nozaki about making sunflowers had set out the
first item on this agenda. But she warned me it would probably take
a decade to prove conclusively what she really believed.

In October 1977, she took out of the library an atlas of molecular
biology. Ever since Watson and Crick established the structure of
the DNA molecule, one of the great thrusts in molecular biology has
been the mapping of the individual amino acids that lie along the
strings of protein molecules. As a result, scientists have been able to
produce maps of protein molecules that look like the itineraries giv-
en motorists by travel associations; a line is drawn from point A to
point B, and the various towns or road junctions or other points of
interest to be found along that line are represented by dots or other
signs. By now, similar linear maps have been produced for thousands of proteins, so that you can start at one end and run your eye
down the length of a protein molecule and see the precise order of
the amino acids—the building blocks of evolution—that make up

the protein. Just as the order of the letters in a word will determine its meaning, e.g., god or dog, so it is the order in which these building blocks are strung down the molecule that determines the biochemical nature of the final protein.

As scientists have studied these maps of molecules, another fact has become clear: Some molecules which evolved in close conjunction in the course of evolution, or which at present share or contribute to the same function, have similar sequences of building blocks along parts of their length. Conversely, if you spot similar sequences of amino acids in two molecules, this might indicate a possible evolutionary connection, if not a shared function.

Anna was fascinated by this biological map-reading. One day, despondent and grumpy, she had relieved her tension by drawing alongside one another the molecular maps of those protein molecules which most interested her: her favorite iron-binding proteins, for a start; certain other iron proteins obtained from plants, called *ferrodoxins;* and also those produced by the histocompatibility complex on the surface of lymphocytes. That was when she was beginning to think that perhaps the iron receptor *was* the same as or close to the complex of surface proteins produced by the histocompatibility gene and had offered to buy champagne should she find this to be so.

This complex is *the* fashionable complex in immunology. As we have seen, the more similar the proteins produced by the histocompatibility gene complex on the surface of the lymphocytes of donor and recipient, the less likely that a transplant will be rejected. To find out whether two people are immunologically compatible, it is necessary to do a crucial test: the Mixed Lymphocyte Reaction test, which reveals the degree of immunological compatibility by detecting the lymphocytes' reactions to each other's surface proteins.

In this test, lymphocytes taken from two people are mixed by a technician; if their proteins are indeed different, the two batches of lymphocytes, "in horror of their differences," as Anna puts it, "start dividing like mad."

Individuals can thus be tissue-typed, just as they can be blood-typed, and in each case the typing identifies the specific assortment of proteins particular to their blood cells or their lymphocytes. The tissue types of all human beings fall into four groups, A through D, though groups A, B, and C include subdivisions.

Anna's thought was that these histocompatibility proteins on the surface of the lymphocytes are themselves, if not the receptor that embraces iron, at least closely associated or involved with it. Nozaki initially undertook two preliminary experiments to test this. First, he tested the effect of iron and transferrin and found that the iron directly inhibited the making of "sunflowers"—i.e., prevented the binding of the sheep red cells to the lymphocytes. It did so, apparently, by masking the lymphocytes' outer surface, as if the iron were blunting the teeth of the body's best guard dogs. Second, he tried the Mixed Lymphocyte Reaction test, mixing two batches of lymphocytes pretreated with iron and also mixing two batches left untreated as controls. And he found that the iron seemed also to prevent the Mixed Lymphocyte Reaction. Instead of the two batches of pretreated lymphocytes' "recognizing each other's differences and dividing frantically," they remained passive, apparently because the iron blindfolded the lymphocytes' "recognition" receptor, and the lymphocytes could not "perceive" another cell's foreignness.

The simple conclusion—that iron might interfere with the surface proteins and therefore with the lymphocytes' capacity to function—had, of course, to be properly confirmed and seriously developed. Furthermore, to do conclusive experiments involving the Mixed Lymphocyte Reaction it was essential to work with someone who knew the techniques of the test very well. Nozaki therefore began to collaborate with Fulton, a postdoctoral fellow in the tissue-type laboratory, the section which specializes in all problems of immunogenetics and the histocompatibility complex. Its work and that of its head are internationally known and highly regarded, and it was natural for Anna to turn to them. When she approached them, Fulton had been at the institute for only three months. He had come to get experience in mixed lymphocyte tests and other techniques. A gentle Georgian, twenty-six years old, with a soft southern voice and exquisite manners, he was really at the beginning of his scientific career. He already had his Ph.D. in immunogenetics from a college in Texas and subsequently had gone through a whole range of veterinarian research experience. From cows to buffaloes to beefaloes—the offspring of a genetic cross between the two animals—Fulton had learned many of the marker tests which are applied to blood samples of animals, to show whether any given beefalo is one-quarter cow and three-quarters buffalo, or some other proportion.

When Fulton and the head of his section met with Nozaki and Anna, she explained her interest in discovering if the surface properties of a lymphocyte could be affected by iron treatment. She went through the logic of her past thinking and pointed out that observation of the amino-acid sequences of certain iron proteins had revealed certain sequences also found in proteins produced by the HLA complex. While there were certainly not so many as to suggest that these proteins were identical, there were enough to suggest that in evolutionary terms they *might* have something in common.

The head of the tissue-typing lab was somewhat less than enthusiastic. He was not convinced by the mapping similarities or by the preliminary data. He couldn't really see the point of the experiments Anna proposed. But he said they would collaborate anyway.

"There you have it," she was to say later. "The crucial difference is between a scientist who thinks something is crazy and yet is willing to have a go, and one who thinks something is crazy and won't even try."

The work got under way about the same time as Edward and Michael's joint venture, but unlike that collaboration, the first few months were very difficult, the technical problems irritating, and the results neither encouraging nor clear-cut enough to show who or what was crazy. The technique for the mixed lymphocyte test is not simple, though in theory it is straightforward enough. The routine proceeds through several complicated stages, each employing different techniques, after which the results must be examined and analyzed. It adds up to about a week for each experiment.

For a while, because the results were thoroughly ambiguous, other people remained very cautious and critical. To an extent, their objections recapitulated those Anna had faced at the Oxford conference: Was the iron really having an effect in this lymphocyte reaction; if so, was the effect due to the iron being toxic; how precise was the mixed lymphocyte reaction; were there biochemical changes in the cultures themselves that were affecting the cells and hence destroying the validity of the results?

In late March 1978, at a meeting to hash over such problems, Anna was heartily cheerful; Nozaki thoroughly glum; Fulton in real pain. He had just given bone marrow for leukemic children, and his behind was very sore. Cheerful or not, Anna reported that other people were much concerned that all kinds of nonspecific effects might be taking place, and she confessed that this worried her so much that she wanted the whole series of experiments repeated.

"It is ridiculous," Nozaki protested. They had done the series with the most scrupulous care.

But Anna insisted. "If you want the others to understand, you have to put up with a lot that is ridiculous. They are critical of *everything* they can think of. You must sense their criticisms and meet them beforehand."

"I don't deny the importance of what you are saying," said Nozaki, "but it is a little bit ridiculous still."

"Never mind," said Anna. "Think up all the criticisms that might be important to other people, and go out and meet them. People are asking whether you are, or are not, getting a genuine reaction. Fulton," she said, "I am not at this bench. You can feel it. Your hands are on the bench, and only the person at the bench can sense whether it is real or not. Is it real?"

"It's real, all right," said Fulton.

"Then," said Anna, "you must think of all the other problems that are important. As you do the work, they will come to you as questions. I can't sense them all."

"Is it something like the old-time aviators flying by the seat of their pants?" I asked. "You can tell when something is wrong?"

"Yes," said Anna. "It is exactly like that."

But Nozaki was still looking unhappy.

Anna laughed. "Nozaki, are you happy?"

Nozaki responded glumly. "Yes, I'm happy. Oh, yes, I'm happy!" he repeated unconvincingly. "But, oh, it is such a lot of work to do all over again."

"From my experience," said Anna, to console him, "it is enormously difficult trying to convince other people—to make them understand—because what they are facing is so new. When you face something new you criticize what you can: method, presentation, even language. This is the problem you have with people in other fields. So *whatever* ridiculous questions they raise, you will *have* to know the answer to them."

Someone had suggested that the results of the experiments had been affected by the presence of heparin, which had been added to the blood samples to stop the blood from clotting.

Anna, in fact, had never used it in any of her own experiments, because it affects the lymphocytes so that they fail to circulate. She said to Nozaki, "You'll have to shake the blood with beads and marbles and stop the clotting that way."

"It takes such a long time," said Nozaki sadly.

When Anna and Fulton had left, Nozaki began to pull folders out of his files and collect his apparatus yet again.

"Oh," he said, with the unhappiness of a man who cannot go straight to his target, "persuasion is rather difficult, and some people are rather obstinate."

When it comes to persuading people, nothing works so convincingly as solid data, and during 1978 the two young men accumulated some very solid data indeed—real nuggets of empirical information. The technical difficulties were gradually reduced in number and finally eliminated; procedures which had been criticized, such as the use of heparin, were either changed or so sharpened that the criticisms were no longer valid; and a whole new range of experiments was set in motion. In the Mixed Lymphocyte Reaction test a batch of lymphocytes was checked not just against the cells of *one* other person but against cells pooled from *four* other people. The idea in this experiment was to confront the responder cells with as many foreign antigens as possible. Thus, from one straight test came twenty-five results. In addition, the lymphocytes were divided into batches of six; five batches were pretreated with five different concentrations of ferric citrate, and one was left as a control. That meant that from each experiment one hundred and fifty results had to be analyzed and the blood from sixty people was subjected to testing, sometimes more than once.

After the first few months of experiment and analysis, they were not at all sure what they had. Some people's lymphocytes seemed to be affected by the iron; others were apparently not susceptible to it, and those unaffected generally fell into one tissue-type group, A-2. Though at one concentration of iron the lymphocytes of A-2 people seemed obviously different in their reactions to treatment, when the data from *all* the groups was subjected to statistical scrutiny the differences became blurred, indicating that these apparent differences were not really significant.

Nevertheless, in view of other experimental evidence, the differences still continued to be suggestive and worth exploring. Moreover, their ideas seemed to be tying in with those of other scientists.

Crucial observations had come from two separate groups, in France and in England. In a disease called idiopathic hemochromatosis, which is known to be caused by excess absorption of iron into the system, iron is stored in great quantity in several organs—usually the liver, although the pancreas can be involved, too, or the heart. The overt symptoms depend on where the excess iron is finally stored. The scientists in England and France had shown there is a significantly greater incidence of this disease in patients with one particular HLA tissue type, the A-3 type. Fulton and Nozaki suddenly found that analysis of their own data seemed to indicate that the *same* tissue-type group was more affected by iron treatment. Anna was not the slightest bit surprised by this, but she was just as excited about it as the two younger scientists.

These observations reinforced her conviction that her general thrust was solid enough and the clues worth pursuing. Now they really were breaking new ground. Nozaki and Fulton were therefore given the go-ahead to do family studies and to examine Anna's questions with the thoroughness that the tissue-typing lab applies to all its human genetic work. They ran the experiments all over again on families, some predominantly of the A-3 tissue-type group, others A-1 or A-2. Then they started on the B and C groups. As Anna said, "We accumulated data like nobody's business."

The science of immunogenetics has linked certain diseases to certain genes in a one-to-one correspondence; idiopathic hemochromatosis and the A-3 gene is one example. One gene–one disease, if not dogma, is at least apparently the current mode of thought. But when Fulton and Nozaki analyzed their new data, no such clear correspondence appeared. Nevertheless, the data derived from twenty-five regular blood donors at the institute proved intriguing enough for the head of the laboratory to insist that they repeat the tests on thirty-five randomly chosen people.

During the weeks and months these experiments were running, Anna was more than preoccupied, her mood more than determined. "There's no room for politeness now," she said, "and there's no room for fooling around. The two or three experiments that Nozaki and Fulton are doing call for an absolute degree of sharpness. They must be impeccable. We are working at the cutting edge of my ideas, and nothing but the finest discrimination will do for convincing others."

In the end, a total of sixty people had yielded nine thousand pieces of information which were reduced to graphs, and these in turn were stared at and worried over.

One Friday afternoon in January 1979, that is precisely what Fulton was doing—worrying because the data on people's susceptibility to iron was not exactly matching the tissue-type groups. On the wall of the laboratory across from him a colleague had pinned a diagram of human cross-reactive groups, showing the classes of blood proteins that react with each other.

As Fulton stared at the chart, looking for another way to analyze the data, he suddenly thought of trying the cross-reactive groups. As soon as this was done, a pattern at once emerged: The lymphocytes of one cross-reactive group were not affected by the iron treatment but others were. Furthermore, the three cross-reactive groups of the B genes showed statistically significant differences in their responses to iron treatment.

To the outside observer, things now appeared to be running smoothly. It seemed only a matter of time before the results could be consolidated and the work would yet again fan out into other areas, while the finer details of existing ideas and experiments would be refined. Thus they seemed to me to be on the home-stretch; it looked like an easy—a relatively easy—canter from then on. I can only describe what actually followed by saying that someone moved the winning post several miles farther on. There was genuine disagreement within the collaborative group as to what the results were demonstrating and whether there was any evidence at all for the links and associations that were hypotheses in Anna's mind. So it was once more into the starting gate, once more into the familiar routine: redoing experiments, rechecking results, refining procedures into more sophisticated forms in order to clear up empirical and conceptual ambiguities.

To get one's colleagues to accept any new concept is probably one of the most Herculean tasks in the whole of science. In this case not only was the whole notion of iron and the vital HLA complex altogether new, almost aberrant in its novelty, but the initial results challenged the widely accepted one-to-one correspondence of one gene–one disease. This the collaborating scientists in the tissue-type lab could not—would not—accept.

Nozaki and Fulton could not then know that this was the low

point in all the twenty-one months that it took before their results were accepted. They were very gloomy. The apparent setbacks and atmosphere of disbelief bore particularly hard on Fulton. This was his first excursion into serious competitive basic research. He did not know if he had what it took to become a serious scientist. Lacking both scientific experience and personal chutzpah (the former was to be remedied), he was unable to argue fiercely with his laboratory head about the validity of the correlation they had demonstrated, let alone its theoretical significance. His fellowship was running out. His future looked bleak.

A year later, in February 1980, I caught up again with the two young men. Nozaki's grin stretched from ear to ear and Fulton was ten feet tall. He had just returned from an International Workshop on Tissue Typing, where he had presented his results. There were no jeers, groans, or indifference now. Many participants in the workshop had been interested and intrigued. Moreover, a scientific paper with all their results was on its way to press, representing his first major contribution in science, and so another milestone in his career would soon be passed.

"Part of the problem earlier," he told me, "was our inexperience in handling this kind of data. I talked it over and over and over with the senior people of my section, trying to figure out what steps to take next. Then I was told to go out, find thirty people at complete random—off the street, as it were—and run the tests on them, before we knew what their tissue type was. So we did. Luckily, the variation of tissue-type groups in this random sample turned out to be exactly the same as in a large population."

The distribution of the tissue-type groups within populations had been worked out some years before. As he explained, "Every time there's one of these big international workshops—once every five years—thousands of scientists are tissue-typed. And so we have a picture of the statistical spread of these groups through populations and among ethnic groups.

"Anyway, we got the results of these new tests on thirty people taken randomly, and again the pattern was just the same. We couldn't make a correlation with resistance to iron treatment with one tissue-type group, but only with a cross-reactive group.

"And we weren't inclined to go ahead with that because my seniors kept insisting that histocompatibility data are always analyzed with a view to the association between one gene–one group–and one

disease, or one characteristic: in this case, being affected by treatment with iron."

"So what did you do?" I asked.

"We just started all over again," he said simply. "We went back
over every bit of the data we had accumulated by then, very, very
carefully. Eventually we put it on computer cards—a technique
that hadn't been available to us earlier. This was my first real experience in being forced to analyze data, and it took an incredible
amount of time to go through it conscientiously and analyze it carefully. It took weeks and weeks. We had to see that the data were in
correct order—were not mixed up. We had to find the very first test
on each person and use only *those* figures, in case repeated bleed-
ings might have had an effect on their response. But finally, when it
was all ready, properly organized and programmed and the computer ran, we found that one group, A-2, kept popping up on the
chi-square analysis."

"The what?" I interrupted.

"The chi-square analysis—a statistical way to compare one antigen with one disease or with one characteristic. This one group kept
showing. In this case, the analysis revealed that in all tissue-type
groups except A-2, the Mixed Lymphocyte Reaction could be affected.

"So it's taken us nearly two years. It was only in October 1979
that my seniors accepted these results, became a little enthusiastic
even. That was when I gave a talk to the whole section."

"How did the workshop go?"

"Very well. People were really fascinated, because it's the *first*
time that anyone has shown that the expression of a histocompatibility protein in *normal* lymphocytes can be regulated by iron. We
know the A-3 genes are *abnormal*. But the really tremendous part
was the hour I spent going over the results, table by table, graph by
graph, with one of the great men in the field. Though he wasn't
jumping up and down, neither did he find anything really critical to
say. Finally he actually said he'd like to do some collaborative work
with us, and that's marvelous."

"Anna offered to buy champagne," I said finally, "if it turned
out that the histocompatibility surface complex was the place where
iron binds to the lymphocyte, or even next door to it. Will there be
champagne?"

"No, ma'am," Fulton replied firmly. "This work hasn't told us about *that*. All we know is that T-lymphocytes in some people, A-2, are resistant to iron treatment so far as the expression of their tissue-type group is concerned, while other groups are affected. But this has told us nothing in detail about the receptor for iron, nor about the mechanism of the effect we're getting. In all the other groups—those that *are* affected by iron—does the iron just sit on the receptor, or does it go in and so affect the biochemistry of the cell that the expression of the histocompatibility complex is altered? We don't know."

I turned to Nozaki. "All these months you've been working on the sunflower experiments, too?"

"Yes, and Anna and I have just published a paper."

Nozaki's work had been beautiful in its simplicity and economy. During the months since he arrived he had been able to vary the lymphocytes' capacity to "make sunflowers" with sheep red blood cells by changing the environmental conditions of the culture and by using iron salts to make those changes. In a gem of an experiment he had modulated this capacity at will: With high iron in his culture, he stopped the T-lymphocytes from "making sunflowers"; in a medium with very little iron, the T-lymphocytes would once more bind to the sheep red blood cells; back again into a high iron medium and they would not. Up and down, in and out, for about five days; and each time the cells changed their capacity to function. The design of this experiment was not new. It was based on one described in a classic paper from the bacterial field reporting experiments that demonstrated how iron can control the synthesis of a bacterium's surface protein. Anna and Nozaki had simply applied the same technique to mammalian cells and were able to present evidence that here, too, iron regulated the expression—the presence or the functioning—of a surface receptor of a cell of the immune system.

"But have you answered the question that was puzzling you when you came to America: Just why do human T-lymphocytes react with sheep red cells in this way?" I asked.

"No, I haven't," he replied. "No one knows why. This marker test has been in use for ten years now, and still we don't know."

"That's all your fault," I teased. "You should have accepted your professor's invitation and taken blood from a lion, a rhino, or a giraffe."

Fulton was curious; he had never heard the story. But as Nozaki wound up with "It is too dangerous," Fulton said with gentle scorn, "That's nothing. I once had to take blood from a buffalo, and they're the most brutal animals when cornered."

"And how," I asked, "do you take blood from a buffalo?"

"Very, very carefully."

"It seems impossible," Anna said to me shortly after, "that only three and a half years have passed since I came to America and we began to fill up the empty space that was to be our lab." In that short time not only had her earlier predictions stood firm, even under the weight of her own and other scientists' scrutiny, but solid additional gains could be chalked up: They had altered a surface property of a normal lymphocyte with iron; they had identified lactoferrin as an important factor missing from the blood of leukemia patients, and they had specified its function. And in the summer of 1980, they reported another find.

In this they were joined for a few months by a German colleague, Dr. Dorner, who two years earlier, learning of their interest in ferritin, had visited them for a brief period. It was the information she had brought along then that triggered a possible explanation in Anna's mind for "making sunflowers," which in turn had led to her famous "Charlie's Angel" letter to Nozaki.

Now Dr. Dorner had returned, bringing with her precious and very specific antisera prepared in her European lab. Sterling work by Edward and Joe—which for Joe had entailed shivering hours of biochemical analysis in the cold room, where freezing temperatures slow down enzyme reactions—had led to the identification of a substance in the blood of leukemia patients which *directly* inhibits the production of normal, but not leukemic, stem cells in the bone marrow.

With Dr. Dorner's help and material they were now able to identify this substance as a very specific form of ferritin; with the extra skill she was able to provide they were also to discover that normal and Hodgkin's disease lymphocytes actually contribute to the synthesis of that particular form of ferritin.

To Edward the discovery of this new factor in the blood of leukemic patients promised to be of even greater potential significance than their earlier identification of lactoferrin and its role in the normal control of white blood cell production. But Anna saw it simply as one more magnificent solid stone in the construction of her edifice, which seemed to gain more strength and beauty as the weeks passed.

Furthermore, confirmation and the confidence it bred were coming from other sources. By mid-1980 the institute's very large computer had been programmed for all the Hodgkin's disease data. Into its maw one could feed almost unlimited quantities of empirical information, whereupon the obliging beast would then compute the most complex correlations and disgorge printouts of its findings; it would even draw the graphs which revealed the pattern and significance of those findings.

One evening in mid-July 1980, Anna sat me down and showed me what the computer had done with the mass of material she and Dr. Ruth had studied four years earlier and with the additional data derived from new Hodgkin's disease patients. (She now had statistics from as many patients as she could wish.) And she grinned.

"It's confirming things I already know," she said, "and telling me others that I didn't know. It's confirming beautifully that the blood serum iron is going up as the white blood cell count goes down . . . the clue that started all this. And it's telling us that there's a direct correlation between transferrin and the dividing activity— the mitogen response—of the lymphocytes. We knew this was true in the laboratory test tube; the computer is showing us that it's true in human beings too. It's telling us a lot."

Looking at the computer sheets and listening to Anna's explanation of them, even an outsider could see that the existence of links between iron and the immune system had been unequivocally confirmed, and that these links could well reveal a complex and vitally important biochemical story. No wonder Anna was grinning. It had been a long, lonely haul to reach this point. Five tiring, sometimes dispiriting, years had taken her along a trail whose first clue had been a pure thought: When lymphocytes appear to have vanished perhaps they are merely hiding. She had begun the journey quite alone, and she and her group had penetrated into new, unpeopled territory. But to one who had followed along with them it was now clear that this empty region was gradually being populated as other colleagues began to enter it from every point of the compass. Together all these people would map the details of the biochemical landscape in the years to come.

"Every time I open an immunological journal," a distinguished immunologist had said to Anna that winter, "I see yet another paper about iron."

Very soon the endeavor would be distributed among many hands, many laboratories, and indeed many subdisciplines of biology, all striving toward a common goal: to determine in the fullest detail the exact role of the iron-binding proteins in the living body, both normally and in pathological conditions. The five arduous years had been a prologue, a good start perhaps, but only the start, of a vast undertaking.

When, in the early summer of 1980, I corraled my friends for the last time, they were all, of course, hard at work and in this sense little had changed. But there were a number of subtle differences. They were more relaxed, so relaxed that they could laugh when I reminded them of their earlier anxieties about the validity of their hypotheses.

They were all deeply immersed in complexity, and biochemical complexity at that, and one by one they drew me into a mire of molecular detail. Individually and collaboratively, each in his or her own particular area of work, bit by bit, process by process, stage by stage, they were trying to put biochemical flesh on the theories. As I floundered and struggled with their recital of laboratory techniques, reactions, chemical questions, and detail, the force of Thorstein Veblen's remark struck home yet again: "The outcome of any serious research can only be to make two questions grow where only one grew before."

"Biochemistry," Anna said, "is now where it's at."

But she herself had changed. She was much older than the person who, only a few years back, had launched into a war dance, saying, "Don't you realize? One day, we shan't have Hodgkin's disease any more." It isn't that she no longer believes in her sanguine forecast; she does, as fervently as ever. But the whole experience of the unremitting collective endeavor, while gloriously rewarding, was also formidably maturing. True, she insists, nowhere but in New York could she have done the work. Her access to the clinical material was extraordinary, and she doubts that clues to iron and the immune system would have revealed themselves to her if she had not been able to study the pathological conditions of human cancer patients.

But it was very rough all round. And she set forth one reason in a letter written on Leap Year Day.

February 29, 1980

I expect you want to know how the work is going. I shall simply say that, on the whole, it is going well from the scientific point of view, generally "Approved but not funded" from the grant point of view (except for Edward's project, and he had a lot of substantiated data preceding our collaboration).

To me, the most rewarding thing is to see that we have a problem to work on. The anticipation—that lymphocytes could do things other than just go round for the detection of specific antigens, and that one possible additional function is protection from the toxicity of metals or the more lethal products of chemical reactions 'encouraged' by metals—this still stands.

Of course we could have found ourselves with no problem to work on, yet this still would have been science as I believe in it, as I learned it, and as I still try to teach it.

But had we no problems, we would be described as "losers." As it is, we are not "winners" yet, but we are threatening "competitors." So, if you asked me what happened to me as a human being doing science in the United States in the last four years, I will tell you, but it will not be very nice.

I lost a kind of naïveté I had when I began. For instance, my reasons for wishing to be anonymous at the beginning of our work together are not the same reasons for wanting us all to be anonymous at the end. At the beginning I wanted us all to be anonymous as a symbolic expression of the fact that, in the long run, what each of us is likely to do will not be remembered as the great individual achievement of John or Mary. Very few names will be remembered in fifty or a hundred years, and probably none in a thousand. But at the end, the reasons for being anonymous lost their naïve and symbolic nature, to become instead part of the awareness that out there, there are no shareholders in this human enterprise—as I had believed. Out there we have "competitors," who might see in our published names the great vile chance for self-aggrandizement. And the introduction of these new values—"competition," "me," "fame," "public image"— into Western science is to a large extent the responsibility of this country. Everywhere in science the talk is of winners, patents, pressures, money, no money, the rat race, the lot: things that are so completely alien to my belief in the way of being human in a world threatened by natural and man-made disasters that I no longer know whether I can be classified as a modern scientist or as an example of a beast on the way to extinction, of little use in these new dimensions of human achievement—as no doubt some great television commentator would put it.

As you see, I became cynical. Very cynical.

I acquired some understanding of lymphocytes and what they do. I lost naïveté, which I used to call purity.

However, for those, the great believers in "fame" and "competition," here is what I think is the most glorious piece of advice. It came from the coach of Heiden, the young skater who won five gold medals in the Winter Olympics at Lake Placid. During the 10,000-meter speed skating event, she kept on telling him, "Run your own race."

And your own race, my dear, is still a human race.

Moreover, he won.

PART III

The flame of conception seems to flare and
go out, leaving a man shaken, and at once
happy and afraid. Everyone knows about
Newton's apple. Charles Darwin said his
Origin of Species flashed complete in one
second, and he spent the rest of his life
backing it up; and the theory of relativity
occurred to Einstein in the time it takes to
clap your hands. This is the greatest
mystery of the human mind—the inductive
leap. Everything falls into place,
irrelevancies relate, dissonances become
harmony, and nonsense wears a crown of
meaning. But the clarifying leap springs
from the rich soil of confusion and the
leaper is not unfamiliar with pain.

JOHN STEINBECK,
Sweet Thursday

CHAPTER 15

THE TWO MARGINS
OF THE RIVER

Time is like a river made up of the events
that happen and its current is strong; no sooner
does anything appear than it is swept away, and another
comes in its place, and will be swept away too.
MARCUS AURELIUS ANTONINUS, Meditations

The source of this book is not to be found in a single spring but in
many rivers of experience. During more than twenty years spent in
the company of scientists, I have been interested in certain relationships particular to this most communal of activities. Of all such relationships, the connection between the individuality of the scientist
and his or her scientific creativity has fascinated me most, and the
problem of characterizing the moments of discovery—the inductive
leaps—has always been the most challenging.

Obviously one cannot grasp these deep characteristics of science
from the outside, either by reading the history of science, by reading contemporary scientific journals, or, for that matter, by reading
academic tomes on the sociology of science. So far as history is concerned, the details of the daily mosaic of science are lost forever unless the scientist keeps a detailed personal and scientific diary—a
habit that seems to have died out in the seventeenth century, the as-
tronomers Kepler and Flamsteed being the last exemplars of this
splendid art. And however charming the personal recollections of
scientists after retirement may be, by then they are often quite un-
reliable historically.

It is the same with the contemporary journals; they provide no
help either. I realized this in September 1965 when Sir Peter
Medawar gave a talk on the BBC Third Programme entitled "Is the
Scientific Paper a Fraud?"* and thus inadvertently provided the

---

*Subsequently published in *The Listener*, September 12, 1963. See also "Correspondence": September 12 and 26, October 10.

impetus for the project of which this book is the result. He was not asking, "Do most scientists cheat?" though some indeed have been known to do so. He was asking whether the writings of scientists, as published in such prestigious journals as *Nature* or *Science*, in any way reflect the human endeavors that underlie scientific invention. To this there is only one answer: a resounding no! For as Medawar reminds us, the scientific paper not only conceals but actually misrepresents the individual human creativity which is its source.

But science is a human activity no less than the arts and humanities whose products are not only more widely accessible but whose human features are much more obvious. The practitioners of science differ widely, and the spectrum of human individuality is just as broad in this profession as in any other. But the most striking feature of the end products of other modes of human creativity is the *expression* of individuality. The music of Mozart is as clearly distinguishable from that of Britten as the paintings of Rembrandt are from those of Picasso. "Look at that," says Gulley Jimson, the hero of Joyce Cary's novel *The Horse's Mouth*, surveying one of his own paintings hanging on the millionaire's wall. "It's unique." Yet this same expression of individual creativity is ruthlessly excised from the final products of science as the profession communally assimilates or rejects one person's ideas. The lasting monuments of a scientific enterprise as garnered by history—the end results which are the permanent evidence of the endeavor—ultimately stand alone, unrecognizable as personal expressions of the human beings who created them.

Yet every scientific creation is no less a unique human event than one of Gulley Jimson's paintings. And since I wanted to know how in science this uniqueness is expressed, there was only one way to find out: Move into the laboratory to observe and try to penetrate the mind of the scientist.

Medawar, who appreciated the problem very distinctly, gave a clear directive: "Peer in through the keyhole." Perhaps he suggested this more distant approach because he could anticipate the strains that might result from the constant contact of observer with observed. It would not be surprising if the scientist came to identify the "science watcher" as Lucy of *Peanuts*, at her most intolerable, saying to Charlie Brown, "Mind if I stick around and watch you for a while? Because sooner or later I reckon you'll do something stupid." Johnson was regularly irritated by Boswell, and from time to time primitive peoples have taken steps to ensure that the anthropologist goes away.

Yet it was in the spirit neither of Lucy nor of Boswell nor of the anthropologist that I embarked on this project, approaching it rather in the spirit of the first historian to realize that the present is history being created and can be recorded as such. I wanted both to record a small piece of scientific history as it happened and also to understand, through the eyes of its creators, how it was happening.

Going into Anna's laboratory for all those years meant taking a deliberate gamble. Discovery can never be guaranteed; nothing might have happened; the record could simply have been an account of one technical or intellectual frustration after another. Nevertheless, I did cherish the hope that by being present over a long period I would be able to catch a crucial moment of discovery, to witness an instant of "the inductive leap," and not only to learn a great deal more about the processes of scientific investigation but to fathom those of scientific imagination.

Even by the middle of 1978 I could sense that the river of thought and work on iron and the immune system and cancer was about to fan into a delta. From that point there were bound to be several streams of simultaneous activity—some were already flowing strongly—and they would continue to widen, converge, separate, and deepen. Many more people would be involved in many places geographically far apart. Thus future developments would be much more difficult to follow and impossible to document in the same detail.

Furthermore, I had never pretended, even to myself, that during these years I was going to be a totally detached observer. Nor could I expect to be present at every moment, at every single activity or meeting. Yet I could hardly have been more involved: through daily communication, whether face to face or by telephone or letter, and also through frequent participation in the laboratory procedures and discussions, all captured by tape recorder or pencil. Indeed, I became so involved that I often forgot my chosen role until, so to speak, I glanced back and caught myself sitting on my own shoulder, frantically taking notes. But the time came—again, in mid-1978—when I began to feel a fierce pride of possession, wanting *these* experiments to succeed and *these* theories to be right. Realizing that my friends the scientists could never allow themselves to reveal such a passion, even if they should share it, and certainly could never permit it to influence what they observe or deduce, I had to accept my being less scientific than they. Certainly I was less self-disciplined. This marked the moment to begin to shift gears.

At the start, four years had seemed a reasonable time to devote to the enterprise, and at the end of those four years I was feeling pretty good about the result. It was therefore a considerable shock when in 1979, as this book was nearing completion, Anna said to me, "You know you missed an awful lot."

The flat assertion suddenly brought to a head the strains of the last five years: the single-minded concentration I needed if I was to absorb a totally new and highly technical subject; continuing and expanding lines of experimental research; the personalities of a new group of people and the evolving patterns of someone else's thought, which tended to crystallize at moments I could never anticipate. Controlling myself with great effort, I asked just what it was I had missed.

"There were lots of things. Things I never told you and things you did not ask, and things I did not tell you just because you did not ask. You sometimes asked about the design of experiments, but you did not ask how I got to the point of designing the experiments one way rather than another.

"Then you never asked me why I had such crazy ideas. For instance, I told you the other day that I have a hunch that lymphocytes could participate in those respiration processes that do not require oxygen. This is quite a statement, you know. It could be either completely crazy or completely reasonable. So I am either mad or I have some reason to say it. I don't think it is important to say the reason at this moment. But you don't say, 'Why the hell do you say a crazy thing like that?' I've only mentioned it to one other person, and this colleague put his reaction in a letter—that it was a bit 'farfetched,' or words to that effect. People react in the same way they react to the telephone when they don't want to be disturbed. They instantly say no or they say absolutely nothing. But nobody actually says, 'Why on earth do you say such a crazy thing?' "

"Well," I objected, "if other scientists don't ask you that question, why should you expect me to? But now that you insist, why *did* you say such a crazy thing?"

"It would take hours—days—to tell. There is a tremendous amount of thinking involved in coming to that conclusion: thinking that goes into reading, and thinking that goes into imagination. And really, you know, you didn't get near the intricacies and detail of what I *think*, and how I came to think of it, and if you didn't how can others? Partly, I suppose, it's because things happen so very rapidly. You have an idea in the morning and you change it in the afternoon because by then you've done an experiment. But, to be fair, there is also the fact that I tend to *explain* what I'm thinking much less than anyone else I know. Even to you I didn't explain the full background to the ideas. So I never really told you."

"Well," I said, "I often wondered just what we *were* getting. Short of hanging a microphone around your neck every minute of the day and night, I can't think what else to have done. There were always times when you were clearly not willing to be pressed or when, because other people were present, it was impossible to press you. But come back to those crucial weeks in April 1977, when your ideas about the evolutionary role of the immune system in relation to iron all fell into place. For me that was the beginning of your real process of discovery—a process whose end point we don't yet know. We don't even know if it *is* a discovery. What did I miss about that?"

"Nothing. I told you just what I was feeling, as soon as I could, as soon as it occurred."

"In any case," I said, "you were indulging in what Joshua Lederberg calls 'those disreputable moments of fantasy, which are so immediately rationalized that there's no way to reconstruct or re-capture their essence, even thirty seconds afterward, even by the person who's had them.' "

"He's right, and I'll tell you something else. If you had been there, I don't think it would have happened. As far as I'm concerned I have to be totally alone for the process even to begin to occur at all. I couldn't have done it with a Herodotus hanging around. But still you don't ask—and I don't tell you—how I came to have these crazy ideas. Nor do other people."

"But you do have a mind that proceeds in a series of kangaroo hops."

"But at least I *know* the intermediate steps! Perhaps you miss things, or my colleagues miss things, because you dismiss them. It's exactly the same when I was looking at S.M.'s spleen for the auto-fluorescence. If I had been looking for it initially, I would have seen it then, as two weeks later I did see it when I was looking for it. Either I didn't see it the first time, or I did and did not think it important. The process is the same in everyday life. People either don't hear what I'm saying, or do, but miss it because they *dismiss* it."

"Let's go back to the original question," I said. "How did you happen to think that lymphocytes might participate in respiratory processes that do not require oxygen?"

"Well, you have the record of the lymphocytes going up in Hodgkin's disease as the serum iron goes down. Therefore you have the possibility that lymphocytes themselves are connected with iron."

"That was when you said that this was all pointing to something very simple."

"Yes. That really led ultimately to the letter to Nozaki and his sunflower experiments, which all point to the fact that lymphocytes do indeed interact with iron. And I believe that we now have preliminary evidence that that interaction is probably controlled genetically, by the major histocompatibility complex and all that.

"But once I decided that the first event in Hodgkin's disease might be a very simple biochemical one, then we had to look for simple biochemical clues, and the only one we had that remotely approached something biochemical was the pain the patients experienced when drinking alcohol. So then I read a lot about alcohol metabolism and tried to connect it with iron. Thus I learned about iron storage in alcoholism. Then there was the rat paper, where there was a change in that enzyme—ALA synthetase—after alcohol intake, and you have all that on record. Then Michael started to figure out how to look for the enzyme, and he found that it was in the sequence which controlled the production of porphyrins. Since we didn't know how to look for the porphyrins biochemically but had found that one of the easiest ways to trace them was by looking for autofluorescent cells, we started on *that* chase. And—here it comes; listen carefully—when we looked for these autofluorescing cells, we found that they occurred regularly in three places: in the red pulp of the spleen where red cells are broken down; the white pulp of the spleen, but only in the T-lymphocyte area; and in the thymus gland. So you have these cells turning up in one place with red cells and in two places with T-lymphocytes. But the autofluorescence tells us that there's porphyrin in those cells, and porphyrin is a building block of hemoglobin.

"It was at *this* point that I began to make a hypothetical link between T-lymphocytes and hemoglobin synthesis. That is the origin of my thought that lymphocytes too may participate in this synthesis in some way and perhaps help in the making of hemoglobin. But I didn't tell you . . . and you didn't ask."

"You made that link at that point because you were reading, or because you were thinking? And why did you make it?"

"Primarily, because I was seeing, and seeing *creatively*. The way I personally function is always based on seeing things and thinking about them. Either these autofluorescent cells—these porphyrin-containing cells—appeared in both the T-lymphocyte areas of the spleen and thymus as well as in the red area by pure coincidence or *there was a reason*. And since I have a fundamental belief that organized structure does not occur by coincidence, I looked for the reason. *The spleen is not a New York party*; it is not a structure that is loosely associated for a short period of time. The spleen is an organ that has been around for millions of years, and therefore if things are found together there they probably have work to do together. So if cells having to do with building blocks of hemoglobin are found in those places where there are also T-lymphocytes, there's probably a functional reason for this. It makes good biological sense. I think there is a tremendous amount of economy in biological function. But I don't talk about this because I don't want to bore people."

"Next time," I said, "please try boring me. If you kept a diary, or wrote down all these possibilities in your laboratory notebooks as Rosalind Franklin did, all this would not have happened . . . and we might have achieved a complete record. You don't put your thoughts into laboratory notebooks?"

"No. Why should I? I don't have reproducible thoughts—not totally reproducible. You must make very precise notes of your procedures and your results because you must be able to reproduce your results tomorrow, or next week. That's what laboratory notebooks are for. But I don't have the same obligation to write down my thoughts."

"That's true. But meantime can we tie up a couple of loose

ends? Do you now know why in Hodgkin's disease the lymphocyte count goes up just before the serum iron goes down?"

"No, I don't know," she replied. "Clearly, I now suspect that the lymphocytes are responding to some change in the amount of iron released into the blood, and they go in there and they pick it up. But we will have to find that out."

"Why did you choose Hodgkin's disease in the first place?"

"That," she replied, "was one of the real strokes of luck in the whole story, because Hodgkin's disease is possibly the *only* disease where the trapping occurs so precisely and neatly. Initially, of course, I chose it because I was so struck with the low levels of lymphocyte function in those patients. But I could so easily have looked at some other disease and got *nowhere*. And look what happened! Look at the luck of that choice. There was Tien-Chun with her mountain of data on Hodgkin's disease patients, and so it went."

"How would you characterize the spleen now?"

"The spleen is a traffic jam. But there is one other thing. Do you remember when we first met and I couldn't decide whether the problem of lymphocytes' getting trapped in the spleen was due to a membrane defect or to the environment of the spleen? And do you remember, too, those experiments in the very first weeks when there was some odd behavior of B-lymphocytes I couldn't understand? *I should have known it was the environment all along;* the behavior of the B-cells told me that it *had* to be the environment. It was a silly mistake. Perhaps I was too excited. Enthusiasm holds you captive.

"But thank God for scientific amnesia. Reading your book has been a most unhappy experience. It has made me relive my excitement and the futility of that excitement. Just take the little discovery I made in the summer of 1975 before I came here—about B- and T-lymphocytes joining up together. I thought then, and I think now, that it was pretty fundamental. We published it in *Nature* without any trouble. And do you know how many requests I've had for reprints? *One.* So even if I think this new stuff is important, I'm not going to get all excited about it.

"What Nozaki and Fulton are doing is really tremendous, but I'm no longer feeling the excitement. Instead, I feel I am taking part in a competitive game. You *have* to be able to talk to someone you trust without being made to feel a fool or inferior because you haven't got a grant, and yet these crazy years, which were the most scientifically fruitful of my life, are almost totally unfundable as American science is practiced. Under such conditions it is sometimes difficult to remember what science is all about.

"But it is important to remember. Science is about scholarship; it is about love; it is about understanding. And what I want to do in the next five years is to nail this thing down impeccably. . . . I want it to be as complete and economical as a poem."

"Well, what have we really got in this book?" I asked Anna in another retrospective moment. "The first part of a continuing process, a description of one piece of history, a document of a process from which it is just possible to generalize to the extent of saying something about the conditions for *your* making discoveries, but which doesn't apply to anyone else? As for the real essence of scientific creativity, perhaps I haven't grasped it. Anyway, even if I had, it would not be applicable to other people."

"I'd agree with a lot of that. You've got something else, though. You've got the excitement and you've got why the excitement is exciting. When I recall what a mess we were in at times, I realize all over again that things that don't work out—the 'mistakes'—are integral to the whole process. You can't tell that a road will end in a cul-de-sac until you've gone down it, and someone must go down it so that other scientists will know that it *is* a cul-de-sac. The reason why science is so absorbing and exciting is that every new fact may be important, and so every new day may be important. As you go through the process, there is no way you can tell beforehand which fact, or which day, is going to be the golden one."

"Then," I said, "each new day is really a new gift from the gods. One knows afterward what was important and significant only because it was subsequently shown to be that; and a mistake, too, turns out to be nothing more than something that was tried and subsequently proved *not* to be important."

"Exactly. And, you know, you've actually recorded a lot more than you may think. You've got something of the elements that go to make a scientific discovery, and you've got some idea of who, in the last part of the twentieth century, should be brought together for purposes of discovery. If, after reading this book, anyone feels that science can be managed like a business, or plugged into computers, or done in a year, then I give up. It is free-ranging ideas, creative thoughts. They are the essentials that must be recognized. And the next essential is to take those ideas and put them into a working pool of competence, where their value is again recognized and where they inspire commitment. You've seen this work beautifully with Edward and lactoferrin. He is a terrific model—an inspiring model—for a young scientist. He has flair, judgment, technical competence, and commitment. With Fulton you see just the beginning. He's got the commitment and he's having to learn the competence and judgment."

"What are the elements in the judgment?"

"There are two, and both come from experience. The first tells you where to move next, and the second is *technical* judgment. There are some people who have all technique and no thought, and that's terrible. But generally when we start in science we have no technique—only visionary ideas about global experiments with cosmic dimension—but no thought and certainly no technique.

"We are like small children hearing beautiful music for the first time. They go to the piano and wave their hands; they imitate all the movements and then their hands come down and one hears a sequence of crashing discords. The trick is to acquire the technique so that one plays beautiful chords, but to retain the childish naïveté, ambition, and enthusiasm during the years of drudgery that you need both for learning the technique and for the unending repetition of technical detail.

"Most importantly, of course, you must identify with what you're doing. You must identify totally. If you really want to understand about a tumor, you've got to *be* a tumor. The wonderful thing is that if you read the early scientific papers of the Egyptians, and sometimes of the Greeks, you know how they themselves actually felt. But as science developed, that changed. You look at a painting and you know a lot, at least you think you know a lot, about what is going on in the mind of the painter. But look at a paper in *Nature*, and you know nothing."

"Are you satisfied with that?"

"Not a bit. I want to change it. Why do you suppose I agreed to do this book? You would perceive *all* this if people really understood—really understood—what the scientific process is.

"Let me come back to discovery. Do you remember *The Little Prince?* The Little Prince tells the pilot that it won't matter when he dies because not only will *he* laugh, but the stars will too. And when the pilot looks at the stars he will hear the echo of the laughter. Now you can make a contribution in science, and its resonance becomes its principal attribute. Then it doesn't matter if you've made only that one contribution. You can die in peace, because the echoes will remain; the thing is *there*. For example, if you read Peter Medawar's first paper on homograft rejections, its echo stays. But in immunology there are only a few papers of that quality."

"What are the qualities that make that paper echo?"

"Well, thoroughness for a start. It is absolutely thorough about one question. It goes into the analysis of that question in every possible way. It covers all techniques available at that time, and thus it has an element of technical thoroughness which makes it a real beauty, even just technically. Then there is the thoroughness of thought. A great deal of thought has gone into it, based on technical knowledge. In fact the two—the thought and the technique—become so mixed that it is impossible to define the boundary between them. Next, of course, his fineness of mind comes through in the quality of his approach to the questions. And you can pick up the quality of his mind again in his discussions of the problems.

"In fact, scientific papers are not devoid of human touches if you know the code, if you know where to find the various clues. In the introduction you can pick up the writer's position in relation to his question; you pick up his individual judgment in the method he, or she, has chosen; and in the way the results are presented, you pick up the quality of mind; finally, you again pick up the fineness of mind in the discussion sections at the end of the paper, when the writer looks at his own particular problems and places them in perspective—in their relation to the problems and ideas currently in circulation."

"It should be possible, then, to look at a scientific paper and sense something of the individual who wrote it?"

"If you know what to look for, yes."

"Anna," I asked later, "have you got any strong views on what the scientific mind is, what the scientific method is? How does it differ from what we do every day?"

"Not one bit. The principle is universal so I've answered *that* question!"

"And ordinary people are as much scientists as anyone?"

"They probably are if they have the guts to be. It is really a matter of interaction again. You are not a scientist because you were born a scientist. You are a scientist, or a poet, or a writer because you have allowed that vertical line in yourself to develop."

"So what we call the scientific mind, or scientific method, is nothing more than a convenient handle?"

"To open what?"

"To open our understanding of the world. According to today's orthodox version of historical changes in the way philosophers have viewed and interpreted science, the early Greeks, when they spoke of the 'God within,' meant that men could reach out with their minds and achieve an intellectual grasp of the universe; Plato tried to visualize the blueprints that, he believed, underlie the structure of the cosmos; with Bacon, Descartes, and Locke, philosophers moved from trying to find out details about the universe to seeking the laws that govern it. Eventually the process got more and more abstract and refined, until we wondered whether our methods were, in fact, revealing the structure of the universe and the nature of the laws governing it, or merely the structure of our minds, the only instrument we have with which to understand the world. This mirroring of our minds may have been inevitable because the world is silent and cannot tell us what *is*. As Kant said, 'Nature is dumb. So the scientist puts nature on the rack and tortures the answers out of her.' Thus doubt is cast on the certainty of our knowledge."

I submitted this to Anna with considerable self-satisfaction at the neatness of my summing up of the philosophy of science through historical time.

"Oh, what a lot of talk!" she said. "I'm going to put it in a totally different way. Again I say the nearest an ordinary person gets to the essence of the scientific process is when they fall in love. In the *Dr. Zhivago* film there is a most touching scene when they go in that horrible train to somewhere or other. There is this mad poet or philosopher or whatever in the train, saying high-minded things as you've just been doing. And there are all these people on top of one another, and it must have been horrible. The smell alone must have been horrible. And this old couple is sleeping on the floor. They are ugly, ugly and blue-colored. They are just sleeping together—this man and this woman—and they hug one another. That is the essence of the scientific process. In other words, you fall in love with operational nature."

"What is the difference between nature and operational nature?"

"Well, if I'm operating with mice I have mouse cells. That is operational nature. I have a section of nature there in the test tube."

"You fall in love with the object of your curiosity?"

"How can you distinguish even *that?*" she objected. "The very fact that something *became* the object of your curiosity is already *process*. It is a horrible thing to say it in the way you do. Objects of curiosity? I prefer to call them things. You fall in love with a thing. Sometimes you fall in love with a fireplace, or you fall in love with a tree, and I honestly think that this is the nearest ordinary people get to the genuine experience of science . . . of what you so pompously call the scientific mind and the scientific process. I think that Kant analogy—the torturing one—is horrible."

"Why?"

"Because it is still influenced by echoes of the Inquisition. It is like rape. Whereas in science . . . it is like the difference between rape and making love. And in my sense and in any other scientist's sense, science is *much* more gentle. Here is a girl, here is a country boy, and for some unknown reason they are attracted to one another. The attraction is different for each of them, but the interaction is reciprocal."

"Now turn it into science."

"Why of all the people in the world should these two come to love each other? The analogy is the same. Here is a cell. It has been going round all the time, and nobody has taken any notice of it. Suddenly you fall in love with it. Why? You, the scientist, don't know you're falling in love, but suddenly you become attracted to that cell, or to that problem. Then you are going to have to go through an active process in relationship to it, and this leads to discovery. First, there is the building up of the attraction, and the object of your attention eludes you. Then you must try to do things to gain its attention with your concepts.

"The boy keeps giving the girl flowers. I keep inventing more refined concepts. And you must understand—that is why it doesn't really matter whether they are my concepts or Edward's or Nozaki's or anyone's—it's the *concepts* that matter. Even if they are absolutely wrong, *it doesn't matter a bit*. The individual enters into it to a very limited extent. So we try to get better and better concepts, trying to get to know the cell. And finally, there is a moment when the girl recognizes the boy, and no longer eludes him, accepts by going up the hill and really getting on with it and expressing it fully! That is the moment of discovery."

"What in terms of science no longer eludes you?"

"The mystery. When you do an experiment that proves your point. That is the orgasmic moment. It is an exciting and most *intimate* moment. You can get that moment from a graph, or from numbers in a machine, or you can see it under a microscope. So that is why," she went on, "whenever you ask me about science, it is always images of enthusiasm, innocence, freshness, and love that come into my mind."

"Do you then go into a laboratory deliberately to confirm or disprove a hypothesis?"

"Of course not," she replied. "No more than you deliberately go into your study to write a verb. It is never like that. I suppose I made my first discovery by accident—or half by accident—but I doubt I'll ever do that again. Normally one is alone with uncertain ideas, or one idea, and that is wonderful; it is exciting. One can almost anticipate the form of the future data. Then *vroom!* it comes, or it does not come. That's the excitement.

"Of course, turning ideas and hunches into a concept is the most difficult of all, and of course the conception of a concept, or the gestation of an idea, is really a kind of birth. It is the only time when men can share with women an essentially female experience; the only time when a man can experience anything like giving birth. When I said this to a man in England, he was frightened . . . or annoyed. He thought I was saying it for feminist reasons. But I wasn't. The birth of an idea, or the birth of a child, or the birth of the universe are really the only times when there is a big bang. It is always accompanied by a lot of pain."

"It carries its own life within itself?"

"Exactly," said Anna. "Yet the odd thing is that, for me, breaking away from an idea, or a concept that I have created, is not painful. The greatest fun about ideas is to toy with them, play with them, embrace them, divorce them, or abandon them. On the other hand, a decision to leave a *place*, to go forward, carries terrible pain for me. I find it much easier to break away from an idea than from a place or a person. And still there's the binomial in me. I always want to be with people, but at the same time I want always to stand outside. I never really want to be part of them. Most of all I don't want to be a part. I must be alone."

"Why did you reject the Kant analogy so vehemently?" I asked some weeks later.

"Because if you have a unifying idea of life there is no such thing as the *external* world. When I find that a lymphocyte has something on its surface, momentarily I am that lymphocyte. I am not finding out anything about the *external* world. We are all part of nature, and if you externalize man—which is to say yourself— you are still the victim of the Inquisition, although in a different way. I, as a scientist, have the privilege of trying to identify in a conceptual sense, with trees or with insects or with how insects identify with trees. The girl is not the external world of the boy who is wooing her; they are part of the same world, and they experience a moment of love and a moment of conception. Then they may go through a moment of disenchantment. But it doesn't matter, because they have had children by then."

"Then," I said, "what about the problem I started with—individuality as it bears or does not bear on science?"

"You are trying to find something unfindable. You are looking for the infinite. You are trying to isolate that quality as a result of which, if certain people were not in a specific place at a specific time, an event would *not* have occurred. I don't think this happens in science."

"All right," I said. "We are not now talking about the essential act of creation, or the inductive leap, or the imagination that envisions new patterns or exploits new techniques in hitherto unknown, perhaps undefinable, ways. We both know that anything we've so far said about that could apply equally to poetry, music, and art as well as to science. Maybe the challenge is to determine the extent to which the individual *matters* in science and the way in which individuality is *expressed* within it.

"If there had been no Beethoven we might have had other expansions of Haydn's symphonic forms, but we would not have Beethoven's symphonies. Whereas in science, though we might not have had Newton, we would still have the laws of planetary motion, and if we hadn't had Crick and Watson we would nevertheless have the structure of the DNA molecule. And, as you have insisted over and over again, if we didn't have you we would, if your ideas are true, eventually have a theory of iron and the immune system and its relation to Hodgkin's disease. In this sense, individuals do indeed not matter. But still—they do matter all the same, don't they?"

"True. But the only difference between us all, men or women, scientists or artists, is what is at the end of one's fingers: a concept for Newton, a shape for Picasso, a word for you. It is not the *process*, and it is not the *thing*, and it is not the people that account for the difference. The difference inheres in the tools you and your fellows are handling. The ability to handle certain tools was acquired early, and this, I think, reflects how easy it is to acquire. Primitive man was already painting beautiful cave paintings on the walls and was able to handle color and form at a very early stage. But man's capacity to handle abstractions, concepts, and mathematical formulae was acquired much later."

"Do you mean that the only genuine difference between all these human activities is the language in which our common experience of life is expressed?"

"Right," said Anna. "And even the language is affected by the understanding and culture of the community. The problem with science is that nobody has stopped to bother to explain the language. I hope that by the time some people have finished reading your book they, who may never have heard of lymphocytes before, will know basically about lymphocyte traffic and the process by which we come to know. It is a very simple idea, just as, I am sure, every idea in science is basically simple. If you cannot put it in simple terms, it is because it isn't a good idea. What I'm trying to say is that the language can be as exquisite as you want to make it, or as complicated as you want to make it, or as comprehensive as you want to make it. You can even make it economical if you choose. But to the man in the street, listening to Stockhausen or to a scientific paper are just the same. He hears sounds, but he really cannot understand a note. The visual arts and the musical arts all call on our basic senses, and so anyone who has no visual defect or hearing defect can experience *something*. But with certain kinds of abstract art—nonrepresentational art—or serial music or science, just 'anyone' cannot. That's why I say the problem is one of learning the language.

"Give a poem to a peasant who cannot read, and he will turn it upside down and it means nothing. Teach him to read, and the poem can become accessible to him as a work of art. Tell him, when he cannot count, that two and two equals four; it means nothing. Teach him to count and he is very pleased, because he's got four cows, all right, and he can measure the buckets of milk his children drink or that he can sell for money, which he can also count.

"So it is all relative. In society *every* person is capable of understanding art and science; and every person should be made so aware of the collective enterprise that he makes the effort to understand and, beyond the effort to understand, is given an experience that brings out the poet or the scientist in the poorest and the least able of us all."

"In Mrs. Wiggins too," I said.

"Indeed, absolutely in Mrs. Wiggins. I am sure this can be achieved. We have already done it in this business of teaching words. And once you really make people at large understand what the scientist is talking about, science is no longer so different from the arts.

"So here is life, a human enterprise which has a variety of aspects. There are men, women, children, and old people in the world, and stupidly we keep putting them in different drawers; men in a little drawer, women in a little drawer, old people in another, scientists in another, poets in another. It is all a great big mistake.

"But still we have, I admit, your original question: about individuality in science. The other night I was talking to a colleague, and we had once again reached the conclusion that the individual didn't really matter in the development of the enterprise. But then, of course, we asked ourselves—we couldn't help it—What at any particular time makes the enterprise so sadly depleted if a particular individual isn't there?"

"The answer," I said with no hesitation, "is just that: the individual. The uniqueness, the lovableness, the infuriatingness, the heart, the intelligence, the style, the panache, everything that goes to make up a personality. It is the absence of these things that depletes the enterprise when a particular individual is missing."

"And how do you, as the observer, discover those qualities in science?"

"There's no way other than the way I have been doing it."

"But when you talk about individual style in science," Anna continued, "what do *you* mean?"

"I think there are classes of scientists," I replied. "I think people go into science for quite different purposes and different motives, which are reflected in different styles. Some people are algebraically—mathematically—oriented. They have no sense of biological form, but they are superb with formulae and brilliant with abstractions. Others love form and try to capture its essence, but not necessarily in mathematical terms. Others can't articulate at all precisely what they are doing, but they are very good at comparison, so they explain important facts with superb analogies.

"There are also the creative manipulators—and I don't mean the technologists or the applied scientists. I mean those who like to play, literally like to play. They like to build, and that's how their ideas may come. Take Rollin Hotchkiss. He would come into the laboratory with a new, extremely simple toy almost every day and play around with it. It was more than a release. It engendered solutions. He once showed me with delight a plate of bacteria which, by division and growth, had made the form of the Japanese ideograph for 'beautiful.'

"Then there are those who don't bother with profound theories and don't miss them either, like electron microscopists. These people are artists of camera and structure. At the opposite pole are those who are fundamentally theoretical *and* philosophical, and who find the whole meaning of the enterprise in the theory. And finally there are the truly superlative people, who are like the tree frog that moves easily between water and land. A first-rate biologist moves easily from the realm of the theoretical into the realm of the experimental. The experimenter in him tells the theoretician what is; the theoretician in him tells the experimenter what they may expect to find.

"I suspect that much of the rest is accidental, except that we are all products of our particular culture. Each of us will see things from a particular stance which reflects the way we were brought up to see things. I also think that scientists establish a kind of personal strategy which is partly unconscious. I've seen it in you when you face a problem: You know what your purpose is and you have your own way of going at the problem. Individuality comes to bear here and again in the choice of the particular road one goes down, or in *choosing not to go down it at all*. The initial strategy of one's experiments may have already been determined; it generally is when one starts as a scientific apprentice. But *you* rejected the preselected strategy and struck out on your own. And so do others, always risking a fearful penalty if they are wrong and almost surely encountering more resistance than if they had stayed within the marked channel.

"And, of course, individuality in science surfaces strikingly in this act of courage. In fact, if you want to make your own pattern of science and have your own 'crazy' view of your field accepted, you can do it. But since you'll be bucking the collective trend, your colleagues will not always appreciate your efforts.

"And, finally, individuality is first expressed in the very act of deciding to become a scientist at all.

"In that respect," I went on to ask, "have you ever felt that science represents an escape, or a release, from human predicaments and pressures? Because in life one seems to pay far more dearly for a wrong hypothesis than one does in science."

"You are absolutely right."

"Is that, in fact, what science does for you?"

"It probably does," Anna replied. "I've never quite looked at it in that light. But it is a marvelous way of looking at it. And I can say this: When you are a highly intelligent person, and you are preoccupied with understanding what is going on, you can always rationalize anything to make it more acceptable to your own critical mind—if you are really clever and want to survive. I suppose the point might have come when I could have said to myself—though I never did—'You are going into science because you are a coward and you don't have the guts to become a real revolutionary or a bourgeois housewife providing grandchildren for your parents.' If I had consciously chosen to be a revolutionary or a housewife, it would have taken a certain courage, and so I could simply have been escaping from being *forced* to act courageously. Because it is too horrible—not to be a housewife, but to 'decide bravely' to be a housewife. And of course I am not going to say I am in science, or I'm doing good science, or trying to do good science, because I am a coward. Because I'm clever I will rationalize, and I will say I am doing good science because I can do *good* science. I am where I can do my best, and so I have everything both ways.

"But do you see why this is so important? If you are really not convinced of your own importance, then you say, 'To hell with my reasons for doing one thing or another. They don't matter.' Whatever the reasons, I can still be that ephemeral bridge, even though the fact may never be acknowledged. I don't think it really matters about the individual, and it is the fate of most of us human beings to be ephemeral bridges anyway. We are back to where we started.

"I still insist that I matter only as a bridge between two collective experiences. I count only if good can come out of a collective experience, whatever that experience is; in other words, if the future can use what has been gained.

"But," she concluded, "going back to your question about escape, thank God for a periodical like *Nature*. Thank God for the marvelous anonymity it provides so we can produce papers that are virtually identical. An anonymous paper in *Nature* is a very good thing for people who would otherwise be incapable of declaiming their results out loud. Thank goodness we are irrelevant to the enterprise, because we can continue to be scientists in spite of our fears. Don't you see? Anonymity is an *achievement*, a real achievement, of the enterprise. It provides a comfortable framework in which we are all *equal* and our agonies are hidden from the public gaze. There are no personal traits being revealed, or insecurities. It is all one security. That is why so many of the people you have met in my world can in fact be very good at science although they may be really frightened in discussions or in meetings. There is no comparable framework for them elsewhere, for the real world is very brutal. So I say thank God for what *Nature* allows—namely, the possibility of not revealing anything of one's self, if one so chooses."

Thus, without knowing it, she pulled me back to that first night in the lab, when, minute after minute, she was performing the ritual, pipetting small amounts of fluid into depressions on a flat dish. When she was tidying up afterward, I had said, "It's a secure world, isn't it?"

"The world of ritual?" she asked.

"Yes."

"Yes, it is."

Anna's New York apartment has two outlooks. The main one overlooks the water, and it is there that she will be most often found, watching the two margins of the river as it anticipates the estuary, her favorite place of all. The current flowing to the sea tracks eastward downriver past the Ambrose light and across the Atlantic, retracing her own path to the Tagus and her origins.

The other outlook reveals twelve windows, six in series, one above the other: the single windows of a research laboratory. There again one can see that science goes on at night, for little winking lights always remain. Peering down I observe the kaleidoscope of activity—arguments, experiments, benchwork, lathework—continuing in patterns now familiar. One scientist has established a comfortable and secure climate in his room with a photograph of a polar bear on the wall opposite the window and some plants on the window sill.

I thought again of Tom Stoppard's statement: "If you can't be a revolutionary you might as well be an artist."

"Or a scientist," I had added. According to Anna, science is egalitarian. Science is, I had seen for myself, all art. Science is, in the true sense of the word, revolutionary, in that it forces us constantly to question ourselves and criticize our ideas and our institutions and thus prohibits complacency. The tragedy for society and science is that we have forgotten the humanity in science, and we have also forgotten, or never learned, how to apply our scientific knowledge humanely. I recalled the conclusion of the letter with which this book began.

If I go on insisting that the individual doesn't matter, except in terms of the collective enterprise of which we are all part, then what this does is to remind us—or should remind us—of the vast anonymous army of people, doing ordinary jobs, of whom no one has ever heard. I matter no more than they do; they matter equally as much as I do. And we are too elitist and proud in the way we have seen science.

Because the permanent God within is the one we' all carry. Within all men and women who, in the end, count. And by count, I mean are capable of moving the whole species—the whole enterprise—forward from where it was when they were born. Within, these people are all rebels: tiresome trades-union leaders, or the Che Guevaras of this world, or those who die anonymous in protests that will never be known. Like Mrs. Wiggins, who never thought to find herself in such revolutionary company, not even on paper.

APPENDIX

During the years that I followed the scientists whose experiences constitute the subject of this book, five new questions were developed. Four were well-delineated and "practical"; one was wider, more theoretical, and just starting to find a focus by the time the book was finished. The clear-cut questions were: 1) Are certain forms of immunodeficiency, such as the one associated with Hodgkin's disease, related to lymphocyte maldistribution? (Chapter 5); 2) Is the interaction of iron with the cells of the immune system regulated by the major histocompatibility complex, generally referred to by the initials MHC? (Chapter 14); 3) Is the iron-binding protein, lactoferrin, a physiological regulator of bone-marrow differentiation? (Chapter 13); 4) Is a particular form of ferritin, present in leukemia, an inhibitor of normal bone-marrow cell differentiation? (Chapter 14).

Whereas in a television interview a question finds its answer within minutes, and in show business or the theater a performance meets with its applause that same night, in science the process of answering questions is much slower and, for the great majority of scientists, the applause, if it comes at all, is rarely audible.

Thus the scientists in this book can consider themselves lucky, for some form of answer to all the stated questions now exists: 1) lymphocyte maldistribution *has* been independently, and directly, confirmed both in studies tracing lymphocytes that can be scanned while "traveling" in the human body, and in other studies of lymphocyte function in the blood and spleen, or blood and joint fluid, etc.; 2) further experiments within their own lab, and in collaboration with those colleagues mentioned in Chapter 14 and others who joined the group since this book was published, have confirmed an association between the response of lymphoid cells and mitogens in the presence of iron and the MHC; 3) the action of lactoferrin as a regulator of myelopoiesis—bone-marrow differentiation—has been disputed, negated, and confirmed by others. Those who have confirmed this action have proposed a model more complex than the one originally arrived at (see diagram, page 182); 4) the answer to this question still waits, since the experimental findings were only recently published.

But what of the wider and more theoretical possibility—the "any old iron" question? This had two aspects: 1) that the immune system evolved having as one main function the protection of the organism from the potential toxicity inherent in the unchecked accumulation of iron resulting from red cell breakdown; 2) that because iron *can* depress immunological functions, and *is* utilized by bacteria and tumors for their growth, it is critical to the balance between organism, infection, and cancer.

These questions, which were becoming scientifically focused as the hardcover edition was being published, are developing along the most interesting and least predictable paths. A number of independent papers have been published confirming the toxicity of iron to cells in the test tube. Recently, a hypothesis has been put forward suggesting that iron accumulation in postmenopausal women could be one cause of the higher incidence of heart disease among this group. In "Dr. Francis's" own lab in England, where he has been using a new kind of very specific antibodies, receptors for iron-transferrin have been identified in culture on the surface of both rapidly dividing lymphocytes and tumor cells. This discovery was made almost simultaneously in three other labs with the aid of new experimental tools. But most encouraging of all is the discovery by another woman scientist in the United States, Dr. Marguerite Kay, showing that the immune system contributes to red cell removal through an antibody against old red cells, which is a physiological antibody against self. Thus, a new function for the immune system does seem to be emerging. Therefore new, unsuspected threats to the organism's health, resulting from high iron intake, are to be suspected.

AUTHOR'S NOTE

This book really is about process rather than discovery. Discovery in science is the culmination of a long series of complicated processes whose end point may not be perceived, or indeed appreciated, by the people who initiate them. In respect to the speculations, concepts, and empirical results that make up the fabric of this book, it must be clearly understood that I do not know—nor does the scientist whose work it records—whether they are in any sense *true*. Nor yet *can* we know. History alone will be the ultimate judge, once the work has been critically assessed by the scientific community— which is currently happening—and incorporated into the body of scientific knowledge—which it may not be. All that I have done here is to record as faithfully as I could one piece of scientific history, without making any value judgments whatsoever as to the science I have been recording or the history I have set down. It is for others to judge whether the science, or the history, is good or bad, spectacular or pedestrian. In this book, it is the process alone that matters.

My insistence on this fact relates to the public exposure this book entails. It is all very well for me: I choose to write books about science and scientists, and I have nothing to lose. But I was asking a scientist to allow me to intrude, to lay bare her activities and thoughts, her half-formed hypotheses, hunches, experimental failures—all those sensitive topics rarely discussed by scientists and then only with friendly colleagues—and to do this over a long period of time. And I proposed to pass on all these to other outsiders and possibly in a form so simplified as to be unfair to her ideas.

There is another problem too. Apart from those few who welcome publicity and often seek it, most scientists are ambivalent about the public communication of science. Though they pay lip service to the need for it, many dislike the notion intensely and almost all instinctively mistrust, on the suspicion of self-aggrandizement, a colleague who gets involved. In the Preface to *Discovery Processes in Modern Biology*,* a fascinating anthology of scientists' personal reminiscences on the human aspects of their work, the editor reports having been warned that the project would "ruin his career," that his status "as a pure scientist would be tarnished . . . as an exploiter," and "you will live to regret this." Thus, though the work recorded here has by now been published in the scientific literature, in order to protect the professional integrity of the scientists, I have, at their request, changed their names. These—and these alone— have been altered. However, my tapes and notes, including the real names of individuals and institutions, have been deposited with the publishers, who have also met the scientists.

I have incurred debts to many people during the past years. First and foremost I thank the scientists for the unique opportunity they gave me and for the delightful times we shared. I also thank the five other scientists and four nonscientists who read the manuscript at a critical stage and gave me opinion, criticism, and advice.

I also acknowledge with gratitude the help and support given by Francis Bennett, Anne Borchardt, Charles Clarke of Hutchinson, and Hilary Rubinstein. I must thank the Rockefeller Foundation for their generosity in awarding me a Rockefeller Foundation Humanities Fellowship which enabled me to work on this project for six months in the first of the five years that it took to complete it, and to The Rockefeller University for hospitality during the whole time. Joy Cull, Elizabeth McCracken, and Carol Zwick patiently typed and retyped the successive drafts, supporting me throughout with skill and humor. Carol O'Brien of Hutchinson added superb critical and editorial judgment as we approached the final stages. But my greatest debt in this regard is to Frances Lindley, my editor at Harper & Row, not only for advice and warm encouragement throughout but for her skill and perception during the editing of the successive drafts of this book. It is with great affection that I thank her.

—JUNE GOODFIELD

* Edited by W. R. Klemm (Huntington, N.Y.: Robert E. Krieger Publishing Co., 1977).